Palgrave Studies in Ethics and Public Policy

Series Editor
Thom Brooks
Durham Law School
Durham University
Durham, UK

Palgrave Studies in Ethics and Public Policy offers an interdisciplinary platform for the highest quality scholarly research exploring the relation between ethics and public policy across a wide range of issues including abortion, climate change, drugs, euthanasia, health care, immigration and terrorism. It will provide an arena to help map the future of both theoretical and practical thinking across a wide range of interdisciplinary areas in Ethics and Public Policy.

More information about this series at
http://www.palgrave.com/gp/series/14631

Alberto Giubilini

The Ethics of
Vaccination

palgrave
macmillan

Alberto Giubilini
Oxford Martin School and Wellcome Centre for Ethics and Humanities
University of Oxford
Oxford, UK

Palgrave Studies in Ethics and Public Policy
ISBN 978-3-030-02067-5   ISBN 978-3-030-02068-2   (eBook)
https://doi.org/10.1007/978-3-030-02068-2

Library of Congress Control Number: 2018960922

This Palgrave Pivot imprint is published by the registered company Springer Nature
Switzerland AG
The registered company address is: Gewerbestrasse 11, 6330 Cham, Switzerland

This book is dedicated to my "partner in crime" (well, that's how she calls me, but being partners involves some form of reciprocity), as well as partner in life, Francesca. Not only is she an incredibly caring, supportive, and smart person, but she is also vaccinated against most infectious diseases. She is my biggest fan and I am her biggest fan, which makes not only work but also, and more importantly, live together extremely funny, happy, and rewarding. Since I wrote this book keeping in mind that she would read it, I have tried to do the best I could, which is more or less what happens with everything else I do in my life.

PREFACE

I have no idea whether, as a child, I got vaccinated against certain infectious diseases. I remember I did catch some of the common ones, and certainly rubella and mumps; for sure, I had not been vaccinated against them. My parents tell me that they did vaccinate me against measles. But I don't know about other vaccines or other diseases. I know I had never been vaccinated against the seasonal flu until a couple of years ago.

But then, a couple of years ago, I started working on the ethics of vaccination and I realized that, for the reasons I will explain in this book, I have an ethical obligation to get the flu jab. Actually, I am quite disappointed that in none of the countries where I have lived so far (Italy, Australia, and the UK), the state has ever *required* me to be vaccinated. As I will explain in this book, states have an ethical obligation to ensure that *all* healthy individuals for whom vaccines are not contraindicated be vaccinated against certain infectious diseases. If this claim sounds too strong to you, I can only invite you to read this book to see whether, by the end of it, you would at least be prepared to consider it reasonable.

Like me, many people of my generation (roughly, those born in the 1980s of the last century), at least from my country (Italy), are likely to be uncertain about their vaccination status with regard to many infectious diseases. This uncertainty is quite telling: when we were kids, infectious diseases, and therefore vaccination against infectious diseases, did not represent a significant concern. As I remember it, in the mind of many people, infectious diseases were not a big deal, and actually they were sometimes welcomed: the worst thing that could happen, according to a widespread conception, was that a child would get the disease, suffer the symptoms for a few days, and fully recover in a week or two, with the benefit of having in the meanwhile acquired immunity against that infectious disease for the rest of their life.

What explains that relaxed attitude towards infectious diseases and vaccines is that back then many people thought that many infectious diseases were relatively harmless. The introduction of some vaccines, and particularly the measles vaccines, a few years earlier had dramatically reduced the number of infections in the developed world; such reduction, in turn, made invisible to many people the possible severe consequences of certain infectious diseases and, accordingly, the benefits of vaccines. For example, as reported by the Oxford Vaccine Group, the year before the measles vaccine was introduced in the UK (1967), there had been 460,407 suspected cases of measles in the country, with 99 measles-related deaths. After the introduction of the vaccine, the number of measles cases per year dropped to around 10,000, with one or two deaths, by the end of the 1980s. Since then, the number dropped further. Vaccines made and are still making a difference. But in a sense, this success backfired: people started to forget, because they could no longer see, that certain infectious diseases can have very severe consequences and even be lethal for certain vulnerable people.

Today, we have easily accessible information about the death toll and the potential complications of many infectious diseases that are likely to be prevented through vaccination. Therefore, it would be relatively easy to see the benefits of vaccines, if only one bothered looking into data provided by reliable sources; access to information could in principle allow people to perceive how beneficial vaccines are even in an era in which the cases of infections and deaths are much rarer than they were in the pre-vaccines era. One would expect that, because such information is available through a simple Google search, the relaxed attitude towards infectious diseases that characterized the 1980s and the 1990s was today only a distant memory. However, reality is very different. Many people and also many institutions still have a too relaxed attitude towards infectious diseases and many have a negative attitude, or at least not a positive one, towards vaccines. We need an "ethics of vaccination" precisely because people do or might fail to get vaccinated for a number of reasons and because states often fail to protect public health through adequate vaccination policies. Too many people failing to be vaccinated pose a serious risk on other people and impose a significant cost on the collective.

Here is an example of the contemporary too relaxed attitude towards infectious diseases. During the 2014–15 Ebola outbreak in Africa, I used to fly quite frequently between Europe and Australia, and I remember Australia had very rigid quarantine measures in place for people coming from overseas and who might have been in some way exposed to the Ebola virus. Every passenger flying to Australia was required to fill in a declaration about their possible exposure to the Ebola virus and about any symptom that might resemble those associated with Ebola. However, I was never requested to provide any certificate of vaccination against common infectious diseases like measles or the seasonal flu, nor would Australia request any person living in its territory to get immunized against such diseases. But diseases like the flu or measles are way more infectious than Ebola, since Ebola is not an airborne disease and is only transmissible through contact with body fluids. Catching Ebola in a developed country like Australia is unlikely, and the risks of sexual transmission is uncertain. Besides, although the death rate of Ebola is very high (on average about 50%) and indeed much higher than that of measles and influenza, these two more common infectious diseases can have very severe consequences and be lethal as well. During 2017, for example, in Europe about 40 persons died because of measles, despite the fact that the measles vaccine is safe, effective, and easily accessible in that part of the world. According to data reported on the Oxford Vaccine Group's website, 250 people *a day* die because of measles worldwide: 1 in every 5000 people infected dies in high-income countries, but as many as 1 in every 100 dies of measles in the poorest regions of the world, not to mention the serious complications of the disease which include, in a country like the UK, encephalitis in 1 in every 1000–2000 cases. Likewise, according to the US Centers for Disease Control and Prevention, the seasonal flu kills between 291,000 and 646,000 people worldwide per year. I am not trying to suggest that Australia (and other countries) overreacted to the Ebola emergency, of course; indeed, I do think that Australia's policy to control Ebola was very appropriate. The point is rather that other infectious diseases would deserve a similar level of attention, especially since they can easily be contained through vaccination without the need for quarantine measures. There is something wrong in being more worried about a disease like Ebola than about way more common and more contagious diseases like measles or the flu: although the death rate of the latter is much lower, their contagiousness can lead to a much higher number of fatalities or severe complications if not kept under control through vaccination. Once again, an "ethics of vaccination" is necessary in order to establish a state's responsibility with regard to the fight against common infectious diseases.

For the sake of clarity, I should specify that there is one aspect of vaccination ethics that I will not mention in this book. Actually, there is more than one ethical issue related to vaccination that I will not address in much detail, and I will explain why in Chap. 1. But there is one particular issue that not only will I not address, but also I will avoid mentioning at all: pharmaceutical companies, or the "Big Pharma", make a profit out of vaccines and therefore have an interest in governments implementing coercive vaccination policies. Of all the arguments you would hear against vaccination or coercive vaccination policies, this is the weakest one, and one that does not require or deserve much philosophical consideration. For this reason, I will only briefly explain in this preface why, despite its popularity (at least within certain circles), I will leave it aside in this book. Coercive vaccination policies are, morally speaking, either right or wrong, justified or not justified. Pharmaceutical companies that produce vaccines certainly make a profit out of coercive vaccination policies. But so what? I suppose that those who appeal to the "Big Pharma" argument picture some sort of conspiracy scenario where governments are lobbied by pharmaceutical companies, or even bribed by them, to introduce coercive vaccination policies in order to pursue the companies' interests. Of course, lobbying and bribery by private for-profit companies are wrong for a number of reasons, and governments should take measures that are in the public interest, instead of in the private interests of a few companies. However, when public interest and private interest of pharmaceutical companies overlap, the fact that private companies profit from the pursuit of the public interest does not matter, morally speaking. Suppose I succeed at demonstrating that states have a moral obligation to implement compulsory vaccination. In this case, the fact that pharmaceutical companies benefit from such policies is irrelevant: good on them and good on us all who will be protected from infectious diseases; in fact, it's a win-win. Moreover, suppose that governments have a moral obligation to implement compulsory vaccination policies, but that the reason why they implement such policies is that they are lobbied or bribed by "Big Pharma"; even in this case, the vaccination policies would not be immoral. We should of course be concerned about lobbying and bribery, but not about the vaccination policy itself; even in this case, the Big Pharma argument is not a good argument against coercive vaccination policies. If, instead, you think that coercive vaccination policies are not morally obligatory or even not morally justified, then it makes little difference whether pharmaceutical companies benefit from them when it comes to ethically assessing vaccination policies; these policies would still be unethical by your ethical standard, regardless of whether companies stand to benefit. The only case in which the profit of pharmaceutical companies matters ethically is the one where coercive vaccination policies are unethical and they are implemented *in order to* promote the private interest of pharmaceutical companies. In any case, since in this book I will argue that coercive vaccination policies are ethically justified and ethically obligatory, the "Big Pharma" argument has no relevance whatsoever for my discussion.

I hope that at the end of the book I will have convinced the reader that the formulation of an ethics of vaccination is necessary and indeed urgently needed. The ethics of vaccination as I understand it here implies that the vast majority of us have a moral obligation to be vaccinated and that our governments have the responsibility to ensure that all of us (with a few exceptions in the case of medical contraindications to vaccines) are vaccinated against certain infectious diseases. In a nutshell, not only is vaccination an individual moral obligation, but failure to vaccinate oneself or one's children should be considered *illegal*. If you think these claims are too extreme, I hope you will find this book, if not convincing, at least challenging. After all, it would be difficult, and I would say even suspicious, to write about the ethics of anything from a philosophical perspective without challenging some beliefs or intuitions.

Oxford, UK   Alberto Giubilini

Acknowledgements

The work on this book was supported by the Oxford Martin School (OMS) at the University of Oxford and the Wellcome Trust. The OMS funded my research within its interdisciplinary project "Collective responsibility for infectious disease" and the Wellcome Trust through the Wellcome Centre for Ethics and Humanities (WEH) recently established at the University of Oxford through the Wellcome Centre Grant 203132/Z/16/Z. This book is made open access in its online version thanks to the generous support of the Wellcome Trust.

I benefitted a lot from the collaboration and constant exchange of ideas with colleagues at both centres. In particular, the regular meetings with the interdisciplinary group at the OMS provided me with an invaluable opportunity to learn about different aspects of vaccination and to test my ideas not only with philosophers but also with social scientists, psychologists, biologists, and historians. The reader will certainly identify within the book the many inputs I got from all disciplines.

Within this group, I would like to thank in particular those people who provided me with very accurate and helpful feedback on parts of this book or who helped me develop some of its central ideas in various ways; in alphabetic order: Thomas Douglas, Sara Loving, Hannah Maslen, Julian Savulescu, and Samantha Vanderslott. Without their help, this book would have contained many more mistakes and inaccuracies than it still probably does. I am also grateful to Andrew Pollard for his help in getting some facts about vaccines right, especially those presented in the first chapter.

Francesca Minerva has also provided me with very helpful feedback after having patiently read the whole manuscript, allowing me to identify and clarify at different points in the manuscript the complicated relation-ship between causal and moral responsibility.

I would like to thank in particular Julian Savulescu—one of the Principal Investigators of the OMS programme and one of the Directors of the WEH—also for having given me the opportunity to work at the University of Oxford and to be based at the Uehiro Centre for Practical Ethics, which he chairs, for the past two years. I know how privileged I have been to be able to work here. The Uehiro Centre for Practical Ethics is, without any doubt, the best place in the world for practical ethics, not only because of the outstanding intellectual and academic qualities of my colleagues, but also for its extremely friendly, welcoming, and stimulating environment, which provides an ideal setting for carrying out work in the best possible way. A condition which, of course, is made possible not only by the amaz-ing academic staff, but also by the incredibly efficient, helpful, and friendly administration staff (Rachel Gaminiratne, Deborah Sheehan, Rocci Wilkinson, and Miriam Wood, in alphabetic order), who allowed me to dedicate basically 100% of my time to research without having to worry too much about administrative issues—something which is also a privilege in academia. My thanks are therefore extended to all these people at the Uehiro Centre who made my life and work in the last two years so easy and interesting.

Thanks also to Adrian Rorheim for the quick and meticulous work of proofreading and formatting the whole manuscript before submission.

Finally, I would like to thank my family for the constant and unqualified support they have always provided me, even when I made choices that, I am sure, they thought were completely mad, such as studying philosophy, moving to the other side of the world, and getting the flu jab every year.

CONTENTS

CHAPTER 1

Vaccination: Facts, Relevant Concepts, and Ethical Challenges

Abstract This first chapter introduces some ethically relevant concepts that illustrate why we need an "ethics of vaccination", such as "herd immunity", "public good", and "vaccine refusal". It argues that the choice whether to vaccinate oneself or one's children is by its own nature an "ethical" choice: it requires individuals to act not only or even not primarily to promote their self-interest but also or even primarily to contribute to an important public good like herd immunity. Besides, since herd immunity is an important public good, ethical questions arise also at the level of state action with regard to the obligations to implement vaccination policies, if necessary coercive ones, that allow to realize herd immunity.

Keywords Vaccination • Herd immunity • Public good • Vaccine refusal • Vaccine delay • Vaccine hesitancy

Why We Need an Ethics of Vaccination

During the 2017–18 flu season, the spotlights of several major Italian newspapers convened on a high school in the Piedmont region. The students as well as all their teachers had decided to get vaccinated en masse against the flu. One might wonder why the newspapers showed interest in such a seemingly insignificant event; after all, many people choose to be vaccinated against the flu every year. What made this particular story

A. Giubilini, *The Ethics of Vaccination*, Palgrave Studies in Ethics and Public Policy,
https://doi.org/10.1007/978-3-030-02068-2_1

1

noteworthy, however, was the reason why the class and the teachers decided to be collectively vaccinated: namely, to protect one of their schoolmates. Some students said they were scared of the needle and of the possible side effects of the vaccine and that they would not have chosen to be vaccinated merely out of their personal desire to be protected from the flu. But one of their schoolmates—Simone—was undergoing cancer therapies and was immunosuppressed at the time, which meant that his immune system was temporarily weakened. Whereas to most healthy people the flu tends to be little more than an uncomfortable inconvenience with few complications, to someone who is immunosuppressed, it is far more disabling and can be life-threatening to a much higher degree. Simone, more than his schoolmates, needed particular protection from the flu.

There are two ways in which an individual can enjoy a relatively high degree of protection from an infectious disease like the flu: one is by being vaccinated and the other is by not being exposed to infected individuals. Unfortunately, according to newspaper reports, Simone could not be vaccinated against the flu because of his weak immune system. I should specify that some details of this story are a bit unclear; in particular, it is not entirely clear whether and why Simone could not be vaccinated: the flu vaccine, unlike some other vaccines that contain weakened forms of the target germ (so-called live attenuated vaccines, or LAVs), is inactivated, that is, it does not contain a live virus. LAVs can be dangerous for immunosuppressed individuals because even the weakened form of a virus could be too strong for their immune system. However, inactivated vaccines are not medically contraindicated for immunosuppressed patients—actually, the inactivated flu vaccine for the immunosuppressed is highly recommended by the medical and scientific community (see, e.g., OVG 2018), considering how dangerous it can be for an immunosuppressed patient to catch the flu. So there seemed to be no medical reason for not vaccinating Simone. In any case, even if Simone could have been—and even if he in fact was—vaccinated, the flu vaccine is less likely to be effective in immunosuppressed individuals. Hence, the only way for Simone to be able to attend school and at the same time remain protected as much as possible against the flu and against its life-threatening complications was to have all his schoolmates and teachers vaccinated as well.

The then Italian Minister of Health, who had been subject to heavy criticisms in the previous months for the new restrictive vaccination policy she had introduced in the country, publicly praised the class' behaviour on social media and paid a visit to the school to personally thank the students. She rightly wanted to give visibility to a behaviour which, she suggested, should serve as a model for others to follow. Many, including all the newspapers that reported the news, had the same reaction as the Italian Minister of Health. In a note on the high school website, the class described their decision to be collectively vaccinated as an "act of solidarity" towards Simone. There is no doubt the class' decision was motivated by noble sentiments and that, considering that many of them would not otherwise have got vaccinated, it was in fact an act of solidarity.

This nice story is particularly suited to introducing a book on the ethics of vaccination for three reasons. First, it clearly illustrates, on a small-scale scenario, the practical application of a concept with great ethical relevance when applied on a large scale, namely, that of *herd immunity*—a concept I will return to later in this chapter and throughout the book. Second, the story shows why we need to develop an "ethics of vaccination", as the title of this book suggests: being vaccinated is a decision that not only could benefit the vaccinated individual but also—and indeed more importantly—contributes to protecting other people around us, thus raising the distinctively ethical question of whether and to what extent we should do something that is not only or even primarily in our self-interest (actually, the individual benefit of vaccination will be minimal or even negligible in some cases, as we will see in Chap. 2). Third, the story suggests that protecting vulnerable people through herd immunity is a collective enterprise, that is, something individuals cannot do alone but need to do together. The collective nature of the effort gives rise to a collective action problem and a tension between collective and individual responsibility. Such tension calls for a philosophical inquiry that can yield precise ethical and, ideally, political prescriptions.

The philosophical inquiry around collective and individual responsibilities will be dealt with in Chap. 2. The policy implications, viewed in light of a principle of least restrictive alternative in public health policy, will be the subject of Chaps. 3 and 4. In this first chapter, I will discuss some of the sources of the ethical problems raised by vaccination and some of the ethically relevant facts about vaccination, clarifying the exact scope of the present discussion and what important ethical issues will be left out.

This book will be successful if, at its conclusion, it will have convinced the reader that in a world where people simply behave in a minimally ethical way—not heroically, only decently—a case like that of the Italian high school class should not be seen as particularly praiseworthy. On the contrary, I hope readers will come to find it quite unnerving that we live in a world where such fulfilments of a basic moral obligation are praised and deemed so special as to be worthy of news coverage. In more specific terms, this book aims to provide a philosophical and ethical framework for conceptualizing and assessing vaccination decisions that supports two theses. First, that being vaccinated is just the fulfilment of a basic moral obligation. Second, that if individuals fail to fulfil this moral obligation, institutions have the moral responsibility to enforce coercive policies to achieve certain public health and social goals.

As I have mentioned above, ethics is, among other things, about whether and under what circumstances we should make choices that are not (only) in our self-interest but also or even primarily in the interest of other people. Unfortunately, the world we currently live in is far from one of moral decency, at least with regard to individual contributions to public health. Widespread lack of morally decent behaviour—that is, behaviour that complies with very basic moral obligations—with regard to vaccination decisions probably explains and perhaps justifies the media attention that the Italian case attracted. Thus, protection of public health through mass vaccination is something that probably requires coercive state interventions. Thus, writing about the ethics of vaccination means not only writing about individual and collective moral obligations but also about the ethical justification for a certain degree of coercion in vaccination policies. The ethical and political discourses are, in fact, not mutually independent; as I will argue in Chap. 2, the individual moral obligation to contribute to herd immunity provides a moral justification for state policies to exert some degree of coercion in order to vaccinate as many people as possible against the most common vaccine-preventable communicable diseases.

I have said above that effective protection of public health *unfortunately* requires some level of state coercion. Obviously, in a perfect world, individuals would contribute to the protection of public health and other worthwhile causes through autonomous decisions, rather than through external impositions; if people behaved morally, coercion would not be necessary. As Angus Dawson observed with regard to vaccination policies, if people were convinced that there is an individual moral obligation to be vaccinated and fulfilled this obligation, compulsory vaccination or other forms of coercion would be unnecessary (Dawson 2011, pp. 150–151). The need for a book on the "ethics of vaccination" stems from the awareness that not enough people are convinced that there is such a moral obligation. Thus, to borrow again Dawson's words, "[r]ather than seeing the justifiability (or not) of compulsion as the central issue in vaccination ethics, we can almost take the fact that this is an issue for public policy as a sign that something has gone wrong with the sense of values in such a population" (Dawson 2011, p. 151).

One might wonder how vaccination could have become such a pressing ethical issue, and why certain policies would even be necessary, given that vaccination is a beneficial medical intervention both for those being vaccinated and for the community at large. Do people not have self-interested reasons for having themselves or their children vaccinated at least against the most common infectious diseases, without having to bring up ethical or other-regarding considerations? Why do people refuse vaccination for themselves or for their children if vaccination is beneficial? These are very reasonable and interesting questions, but they are not the kinds of questions I will primarily aim to answer in this book—although I will try to provide some answers later in this chapter. This book is not primarily about the reasons, the motives, or the sociological explanations for why individuals refuse vaccination for themselves or for their children (about which excellent contributions already exist, such as Largent 2012; Navin 2015), nor is it about what strategies could be effective in *convincing* people that vaccination is the right choice to make. This is a book about what kinds of moral obligations people and institutions have with regard to vaccination, regardless of what psychological, social, or cultural factors prevent them from fulfilling such obligations. It is a book about moral values involved in vaccination decisions, rather than about facts about vaccines and vaccination decisions. But of course, facts and values are closely related in the sense that certain facts about vaccination and vaccination decisions do have ethical relevance, that is, they generate certain moral obligations once we agree upon certain very basic and reasonable ethical principles.

For example, here is a fact about vaccines that matters ethically, in the sense that it generates individual and collective moral obligations: society as well as individuals could experience seriously bad consequences, including death, as a result of vaccine-preventable infectious diseases. In 2017, there has been a fourfold increase of measles cases in Europe, going from slightly more than 5000 cases in 2016 to more than 21,000, and about 40 people died of measles in the same year in the European region (WHO 2018). Keep in mind that we are talking about an area of the world where vaccines are easily accessible and relatively cheap. It is unclear how many of these people (if any) were unsuccessfully vaccinated (after all, the measles vaccine is "only" 93–97% effective, depending on how many doses are administered) or not vaccinated at all against measles, and if so, how many of them had medical reasons for not being vaccinated. It is very plausible to suppose that the vast majority of these cases could have been prevented through vaccination—either of the victims or of the people around of them, or ideally both; as the European Centre for Disease Prevention and Control reports, "of all measles cases reported during the one-year period 1 December 2016 to 30 November 2017 with known vaccination status, 87% were not vaccinated" (ECDC 2018). Since the vaccine against measles—nowadays usually administered together with the mumps and the rubella vaccine in the so-called MMR vaccine—has been around for about 50 years, all the while proving itself to be very safe and effective, one would think that there are more than a few ethical issues raised by vaccine refusal. If these 40 people had been vaccinated, or if they had been successfully protected by herd immunity as a result of those around them having been vaccinated (in the same way as the Italian high school students got vaccinated to protect Simone), these 40 people would probably not have died—I say "probably" because we cannot exclude cases of vaccine failure and low vaccine responders as a possible genetic trait. Therefore, at least some unvaccinated individuals are *causally* responsible for the deaths of these 40 people. But as I will argue in Chap. 2, any non-vaccinated individual, regardless of whether they directly infected other people or not, fails to fulfil their *moral* responsibility to contribute to the prevention of the illnesses and the deaths that occur for vaccine-preventable infectious diseases. Grounding such moral responsibility will require some ethical and philosophical analysis of the concepts of "individual" and "collective" responsibility, which I will undertake in Chap. 2.

Before moving to a more detailed explanation of what an ethics of vaccination is and why it is necessary, three clarifications are in order.

First, when I talk of vaccination, I am not, of course, referring to any possible vaccine available. Certain diseases are not a threat in many parts of the world, particularly Western countries, and there is no need to be vaccinated against those diseases unless one plans to travel in areas of the world where those diseases exist. Examples include vaccines against yellow fever and cholera. This book is not about vaccination ethics for travellers, which is in any case an important and underexplored issue in public health ethics; rather, it is about the ethics of those vaccinations that are typically recommended or mandated in the vaccination schedules of

Western, developed countries. These include the MMR, influenza, pertussis, "6-in-1" (which contains vaccines against six different infectious diseases, including polio), pneumococcal, and rotavirus vaccines (for a list, see, e.g., NHS 2016). Also, as my analysis of the ethics of vaccination unfolds in the next chapters, it will become clear that my arguments only apply to those vaccines that protect against *communicable* infectious disease and therefore not to vaccines against any infectious disease. For example, the ethical considerations I will make do not apply to the vaccine against tetanus, which is not a communicable disease (although the tetanus vaccine is typically administered through the 6-in-1 vaccine, which also contains vaccines against communicable infectious diseases).

Second, I should clarify that when I talk of vaccination, I will refer both to adult and child vaccination. Typically, vaccination targets children of different ages, and even for a vaccine that is commonly chosen by people of all ages, such as the flu vaccine, there are good reasons for vaccination policies to target children rather than adults, given that children suffer higher influenza incidence rates and are therefore more likely to cause seasonal influenza epidemics (Bambery et al. 2017). Thus, vaccination choices are often choices that adults make on behalf of their children. But adult vaccination is equally important from the point of view of public health, given that adults contribute to vaccine coverage rate and to spreading infections in the same way as children do. It might be thought that referring to both types of vaccination at the same time creates problems when it comes to discussing ethical obligations with regard to vaccination; for example, it is one thing to say that an individual has an obligation to be vaccinated, and it may be quite another thing to say that an individual has an obligation to vaccinate a child who is not competent, or in any case does not have the authority, to consent. I will address this concern in Chap. 2, when I discuss the ethical obligations with regard to vaccination decisions.

Third, I will not be talking about special obligations of certain particular groups—for example, health workers—with regard to vaccination. The reason is simple: since I will be arguing that *everybody* (with a few exceptions) has a moral obligation to be vaccinated and should be subject to a legal obligation to be vaccinated, talking about "special" obligations of certain subgroups would not add anything substantial. For instance, health workers have a moral obligation and should be subject to a legal obligation to be vaccinated not *qua* health workers but simply *qua* members of communities with the collective responsibility to realize herd immunity.

The Luxury of Vaccine Refusal and Delay

Although this is meant to be a book about the ethics of vaccination, and not about vaccination facts, it goes without saying that certain facts require some scrutiny if we want to adequately understand the ethical issues they raise. In particular, it is useful to say something about why many people today fail to vaccinate themselves or their children, thus exposing them and others around them to easily preventable infectious diseases or in any case exposing them to infectious diseases for longer than necessary.

Let us start by pointing out that referring to all these people as simply "anti-vaxxers", as many do and as the media usually call them, can be misleading. The term "anti-vaxxers" might be a useful label to indicate very broadly the group of people who, for *whatever* reason, are against vaccination; but it does not do justice to the complexity of reasons or psychological explanations for why people fail to vaccinate themselves or their children. For example, some people who refuse or delay vaccination do not consider themselves to be *against* vaccines as such (as the term "anti-vaxxers" seems to suggest), but rather in favour of "safer" vaccination programmes, thereby excluding some vaccines from the group of the safe ones. Besides, there are different factors, apart from beliefs about vaccine safety and effectiveness, which explain people's opposition to vaccination; below, I will review some of these factors.

Following Mark Navin (2015, p. 2), anti-vaxxers who deny the safety of vaccines can be referred to as "vaccine *denialists*". Not all those who fail to vaccinate are vaccine denialists, though. For one, some of them might fail to vaccinate not because they believe that vaccines are unsafe or ineffective, but because they have moral or religious views that are incompatible with the use of vaccines, or simply because they prefer to free-ride on the protection that a sufficiently high percentage of vaccinated people in the community guarantees through "herd immunity" (a concept to which I will return shortly). Moreover, parents are often "hesitant" about vaccination, rather than outright vaccine denialists. Vaccine "hesitancy" refers to the vaccination attitude of people who do not refuse vaccination in principle and hence are not, strictly speaking, "anti-vaxxers" or vaccine denialists. They simply have concerns about whether vaccines are really safe and/or effective, rather than strong beliefs about safety and effectiveness; or alternatively, they might believe—mistakenly (CDC 2018)—that it can be harmful to administer many vaccines at the same time and thus tend to delay vaccinations or opt only for certain vaccines at any one time (Dubé et al. 2014a).

Of course—and this is a relevant distinction in order to circumscribe the focus of this book—we also need to distinguish non-vaccination that is due to people's choices or negligence more generally (including, as we will see, the negligence of giving in to unconscious biases) and non-vaccination that is due to factors beyond people's control. Sometimes people do not have (easy enough) access to vaccines, particularly in developing countries (Favin et al. 2012), but also in developed ones—especially in those with high rates of immigration. Distance from health facilities, internal population displacements and insecurity, and the fact that many illegal immigrants are afraid of being reported to the police if they visit health facilities (Dubé et al. 2014b) are among the factors that might hinder vaccination uptake in many countries. These circumstances contribute to the spread of infectious diseases as much as, if not more than the sociological, cultural, or psychological factors that influence individuals' choices not to vaccinate themselves or their children where vaccination is easily accessible.

In fact, difficulties in accessing vaccines account for a significant number of cases of preventable diseases and death worldwide. It has been estimated (Durrheim and Crowcroft 2017) that measles vaccination saved 7.1 million lives worldwide between 2000 and 2015. This looks like a remarkable datum, as it obviously is in many respects. However, this figure pales in comparison with the 114,900 people who died of measles worldwide only in 2014 (Perry et al. 2015): if several million lives were saved where vaccines are easily accessible, it is simply unacceptable for 114,900 people to die in one year of the same easily preventable disease just because many of them have difficulties accessing vaccines—just as it is simply unacceptable, to compare, that malnutrition and starvation still exist in certain parts of the world while there is overabundance and waste of food in others. Although these 114,900 deaths represent a stunning 79% decline in measles-related deaths from the 456,800 fatalities of 2000, they remain an objectively too high death toll for a disease that is vaccine-preventable, especially in light of the fact that, since 2010, progress towards the WHO's goal of eliminating measles from four WHO regions has significantly slowed down (Perry et al. 2015, p. 623). The vast majority of those 114,900 deaths are not the result of people's choices, as is likely the case for most if not all of the about 40 deaths of measles in Europe in 2017.

What all this suggests, among other things, is that opposition to vaccines is literally a "first world problem"—not in the trivial everyday sense of the term, of course (quite the opposite!), but in the sense that it is a luxury of people in the first world to be in the position to make the *choice* whether or not to vaccinate oneself or one's children. Granted, opposition to vaccines exists in other parts of the world, too. But death rates in many parts of the developing world are often attributable to access problems, although these may disguise the issue of opposition to vaccines in those countries; on the contrary, the fact that in the developed world we have limited problems of access to vaccines suggests that some form of opposition to vaccines represents the main problem in these areas. The about 40 people who died of measles in Europe in 2017 were the result of people's choices, including the choice not to choose regarding vaccination and to accept the status quo (which, in countries where vaccination is not mandatory or compulsory, is non-vaccination). Thus, being a book about the ethics of vaccination *decisions* and the ethics of whether and how people's decisions ought to be *constrained* through vaccination policies, this might be thought of as a book about an ethics for the privileged. And in fact it is, in the same way as books about the ethics of food propose an ethics for the privileged that are in the position to make choices about which kind of food to consume, for example about whether or not to be vegetarian. A comprehensive ethics of vaccination would ideally include prescriptions about which measures ought to be taken at the international level to address the problem of partial or complete lack of access to vaccines in certain parts of the world and in certain subpopulations within developed countries. This is an important challenge and one that international health agencies—the World Health Organization (WHO) *in primis*—are aware of and are working hard to address. But this book does not have the ambitious purpose of covering all the possible ethical issues raised by vaccination and non-vaccination. Addressing the problem of insufficient access to vaccination requires confronting issues of international politics, including the economic and health aid that developed countries ought to provide to poorer countries, as well as issues about facilitating illegal immigrants' access to healthcare services— after all, the level of public health in a country also depends crucially on the level of health of its immigrants. These considerations, even if not less important than the ones I will be discussing, are beyond the scope of this book. The "ethics of vaccination" will be understood here as the ethics of individual vaccination *decisions* and of vaccination policies that might sway or determine such individual decisions.

Although I have said above that the term "anti-vaxxers" is too broad to capture the complexity of the phenomenon of vaccine refusal, it remains a useful label to refer to those privileged individuals who actively *choose* not to vaccinate themselves or their children for *any* reason. Now, it has been observed that the perception of the impact of the anti-vaxxers on low vaccination rates tends to be greater than it actually is (Kahan 2014). Also, Samantha Vanderslott has pointed out that scepticism about vaccines or even outright opposition to vaccines often does not translate into actual vaccine refusal—a mismatch that in her view is an instance of the more general psychological phenomenon known as *attitude-behaviour gap* (Vanderslott 2017a). At a first glance, these two considerations seem to suggest—as indeed Vanderslott (2017b) has suggested—that the anti-vaxxers' impact on vaccination rates is relatively insignificant. For example, in the US, the median rate of active vaccine refusal in the case of parents of school-age children—that is, refusals by actual anti-vaxxers—is 2% (Seither et al. 2017). Perhaps we should not be too worried about such a small proportion of individuals. If this were true, then an ethics of vaccination decisions or of vaccination policies would not be that important, because enough people would already be convinced that vaccination is the right choice and they would not need to be given further ethical reasons or to be coerced by restrictive vaccination policies. Thus, according to this view, where vaccination rates are not high enough, there probably are other factors—such as difficulties in accessing vaccines—that need to be considered, rather than vaccine denialism or a more general anti-vax sentiment. In this scenario, individual decisions and coercive policies would play a relatively small role in determining vaccination rates.

However, according to Vanderslott, the explanation for the mismatch between widespread anti-vaccine sentiment and not-so-widespread vaccine refusal "varies from social pressure to repercussions for not vaccinating" (Vanderslott 2017b). For example, disagreement between parents about child vaccination typically results in rulings in favour of the pro-vaccination parent; and there are penalties that states impose for non-vaccination which constitute strong disincentives for vaccine refusal (such as preventing school entry to the non-vaccinated, as happens in the US, or withholding certain financial benefits, as happens in Australia). But if this is the account offered to explain the small impact of the anti-vaxxers on vaccination rates, then the explanation is question-begging and raises precisely the ethical issues that this book aims to address. According to this type of explanation, the low rate of active vaccine refusal (e.g., in the form of applications for non-medical exemptions from vaccine mandates in the US) would be due not to the low number of anti-vaxxers, but to external pressures, including how difficult it is to obtain non-medical exemptions and state coercion. Whether such external pressures and state coercion are legitimate is precisely the question that raises the ethical issues that I want to address in this book, namely, whether there is an ethical obligation to vaccinate oneself or one's children and whether a certain degree of coercion, and what degree of coercion precisely, is ethically acceptable or even ethically required in the implementation of vaccination policies.

One important aspect that Vanderslott's reflection raises is that vaccination attitudes must be distinguished, with respect to their practical implications, from actual vaccination decisions. As already said, someone might be deeply opposed to vaccines for a number of possible reasons, but still decide to vaccinate their children for a number of different reasons—including the desire to avoid heavy penalties. Or someone could in principle be in favour of vaccines, or in any case convinced of their overall beneficial effects at the individual and at the collective level, but still decide not to vaccinate themselves or their children, for example because they think that it is safer or more convenient to free-ride on the herd immunity that other members of the community have realized. Now, what matters ethically—or at least this is the stance I will assume in this book—is primarily vaccination *decisions* and only secondarily vaccination *attitudes*. Ethics provides people with certain types of reasons—such as moral obligations—to make certain decisions rather than others. And moral obligations exist regardless of whether people's attitudes align or not with them. As suggested above, it would be ideal if individuals did vaccinate themselves or their children autonomously, because they were convinced of the benefits of vaccines and aware of the fact that vaccination is a moral obligation. However, ultimately, what matters the most is that individuals *do* vaccinate their children, whether or not they think that it is beneficial or morally required. Because vaccination actions matter more in ethical terms than vaccination attitudes, it is important to develop, alongside an ethics of vaccination decisions, an ethics of vaccination policies. Just as ethics in general is about how we should live and what we ought to do, and therefore about how we ought to make practical decisions, so an ethics of vaccination is ultimately about what individuals, collectives, and institutions ought to do with regard to vaccination decisions—that is, about what moral obligations different actors must fulfil. Of course, this is not to say that individual dispositions, beliefs, concerns, and fears do not matter. Indeed, they have great value, both intrinsically and instrumentally: intrinsically, because it matters morally how people feel when they make certain choices rather than others, and it is morally preferable that choosing vaccination did not undermine their psychological well-being; and instrumentally, because correct beliefs and a correct attitude towards vaccination make it more likely that individuals will fulfil their moral and legal obligations to vaccinate. However, these considerations are of secondary importance. Once we have established that there are certain moral obligations to fulfil and that certain legal requirements would be ethically justified, then individuals have those moral obligations and ought to abide by those legal requirements regardless of what their beliefs and attitudes are. Surely we (which is to say governments, public health authorities, and people who have the capacity and power to influence public opinion) ought to do whatever we can to make sure that as many people as possible are well-informed and have the right kinds of attitudes towards vaccines, for example, through adequate information campaigns and by promoting trust relationship between the medical and scientific community on one side and the wider population on the other. But ultimately, whether or not these attempts are successful does not affect the strength of moral obligations and the legitimacy of coercive vaccination policies.

It is, however, interesting to survey the factors motivating the sort of attitudes towards vaccinations that ultimately result in a total or partial failure to vaccinate where vaccines are easily available. As we will see in Chap. 3, understanding how these attitudes originate might be useful in order to design effective vaccination policies. The factors that explain failure to vaccinate can be divided into four types: sociological, epistemic, cultural, and psychological.

The first type of factor—the sociological one—is the most problematic to describe, for the simple reason that it is unclear whether it actually is a factor that determines vaccination attitudes at all. In particular, it has proven quite difficult to draw correlations between socio-economic status and vaccination decisions. For example, in 2014, Wang and colleagues published a systematic review about the socio-economic status of parents who applied for non-medical exemptions from school vaccination requirements in the US, where in most states parents can be exempted from the mandate through "conscientious objection" to vaccination (Clarke et al. 2017; Navin and Largent 2017; Giubilini et al. 2017). Two studies showed that parents requesting non-medical vaccination exemptions in the US tend to be white and college-educated and with a higher income than those who did not seek an exemption; however, two other studies found that parents applying for exemptions are more likely to have lower socioeconomic status and that parents with lower household incomes were more likely to oppose compulsory vaccination than those with higher income (Wang et al. 2014).

The same review also suggested that the belief that vaccines harm the child is a common and persistent concern among parents who seek non-medical vaccine exemptions. This is the epistemic explanation for vaccine refusal or delay. As is easy to imagine, some parents are vaccine denialists at least to some degree, in that they are simply doubtful of the efficacy or safety of vaccines (Smith et al. 2011; Harmsen et al. 2013). Many of them believe that the risk of iatrogenic diseases (i.e., diseases caused by excessive attempts to treat or prevent another medical condition) resulting from vaccination is greater than the risks deriving from the disease that vaccination would prevent, and that therefore it is not worth taking it (Salmon et al. 2005; Wang et al. 2014). Others believe that it is sometimes beneficial to catch an infectious disease because the disease would strengthen the immune system and therefore protect the child from future, and perhaps more severe, diseases (Hough-Telford et al. 2016). All these beliefs are false, at least in most circumstances (as we will see in Chap. 2, when vaccination rates are very high, the first type of belief might be correct). Therefore, the problem here concerns how people come to form certain incorrect beliefs about medical fact; in other words, the explanation for the failure to vaccinate is epistemic in nature.

Some parental opposition to vaccines can, however, be explained by what I have called the cultural factor. In this case, the explanation refers to some ethical or religious aspect of the cultural background of people who refuse or delay vaccines. Some people have ethical reasons for opposing vaccines; for example, some have ethical quandaries about using vaccines that contain viruses grown from cell lines derived from aborted foetuses or animals (Salmon et al. 2005). However, it should be noticed that it is likely that the facts about vaccine manufacture that these people have in mind are ethically less significant than they think. For example, the two only human foetal cell lines used to grow viruses for vaccines today are derived from two foetuses aborted therapeutically—that is, not for the purpose of deriving cell lines—in the 1960s. All the other vaccines that require cell lines derive them from animals, and even among these vaccines,

only four are commonly mandated or recommended in standard vaccination schedules, or are anyway normally administered: the hepatitis A,
rubella, chickenpox, and zoster vaccines. Meanwhile, other people are
opposed to vaccines because they belong to certain religious groups with
specific prohibitions. However, it is worth pointing out that it is difficult
to correctly attribute vaccine refusal to religious beliefs. For example,
while a 2005 survey of parents in the US found that 9% of parents refused
vaccination on the basis of religious beliefs (Salmon et al. 2005), a 2014
WHO report found that, according to a survey among immunization
managers in different countries, religious beliefs were perceived to be the
most common determinant of vaccine hesitancy (WHO 2014). What
accounts for this discrepancy between two different interpretations of the
role of religious beliefs in vaccine refusal? Part of the explanation might be
that religious opposition to vaccines is sometimes misattributed. For
example, it has been suggested that one of the reasons why Amish communities in the US have very low vaccination rates is not, as the myth
goes—and as I have suggested in a previous publication (Giubilini et al.
2018)—that they have a religious opposition to vaccines, but simply that
it is relatively difficult for isolated Amish communities to access vaccination services (Wenger et al. 2011). Besides, even if the phenomenon of
vaccine refusal is quite widespread among some Christian religious groups
(such as Christian Scientists, Dutch Reformed Church members, or the
Amish), it seems that the Catholic social teaching is not incompatible
with, and indeed does entail, a moral obligation to vaccinate in order to
protect the community against serious harm (Carson and Flood 2017).
Therefore, religion might play a more limited role, both psychologically
and philosophically, than commonly thought in an explanation of vaccine
refusal and vaccine delay.

It could reasonably be argued that a similar problem regarding correct
attribution of reasons for vaccine refusal or vaccine delay exists with respect
to any of the self-reported reasons just mentioned. How so? The answer
has to do with the fourth kind of explanation for vaccine delay or refusal I
mentioned above, namely, the psychological explanation. Regardless of
what reasons people provide for their opposition to vaccination, much of
this opposition turns out to be irrational, at least according to a psychological definition of (practical) "rationality", that is, as the capacity to make
decisions based on conscious reasoning rather than merely on unanalysed
intuitions and emotions. According to Joshua Greene's characterization
of rationality, "reasoning, as applied to decision making, involves the

conscious application of decision rules (…). Reasoning frees us from the tyranny of our immediate impulses by allowing us to serve values that are not automatically activated by what's in front of us" (Greene 2013, p. 136). I will accept this psychological definition, whereby a decision is rational if it is based on reasons that the agent is aware of (of course, other, more philosophical notions of "rationality" would not consider this as a sufficient or even a necessary condition for rationality). Now, as it turns out, rationality thus understood is not what many people rely on to make vaccination decisions. Let us analyse the issue of rationality in vaccination decisions in more detail.

If most vaccination decisions were actually based on rationality, it would be difficult to explain why, as Mark Navin has concluded from his analysis of vaccine refusal, many vaccine refusers know more about vaccines than do parents who vaccinate (Navin 2015, p. 10). If vaccination decisions were based on knowledge of facts about vaccination, including their safety and effectiveness, rational people would opt for vaccination in spite of the small risks of iatrogenic diseases involved, at least when vaccination coverage rates are low and protection from infectious disease hence cannot be guaranteed through herd immunity. But the fact that many vaccine refusers or vaccine-hesitant people have fairly good knowledge of vaccines suggests that, in many cases, decisions not to vaccinate are not based on reason alone, at least as defined by Greene.

And in fact, psychological research seems to support the thesis that many decisions to refuse or delay vaccination are of an irrational nature. For example, while public health authorities often encourage doctors to discuss risks and benefits of vaccination with parents who are opposed to vaccines (Omer et al. 2009), some evidence seems to suggest that many sceptical parents are unlikely to be swayed by risk-benefit analysis of vaccination (Meszaros et al. 1996). Further psychological research has suggested that vaccination decisions are often likely to be the result of biased judgements, rather than of cool reasoning. A bias can be defined as an unanalysed prejudice that leads to systematic errors or deviations from rationality standards in judgements or decisions. In particular, psychological research has brought up "omission bias" and "naturalness bias" to explain much of the opposition to vaccines. Omission bias can be defined as "the tendency to see a negative outcome resulting from inaction (omission) as more favourable than the same negative outcome resulting from action (commission)" (Di Bonaventura and Chapman 2008, p. 2). In the case of vaccination, omission bias is the tendency to see the possible negative outcomes resulting from infectious diseases, and hence from non-vaccination, as more favourable than the negative outcomes resulting from vaccination. The naturalness bias is "the tendency to prefer natural products or substances even when they are identical to or worse than synthetic alternatives" (Di Bonaventura and Chapman 2008, p. 2). Now, strictly speaking, it is not correct to consider the vaccines routinely offered or mandated as "synthetic", because these vaccines contain the very same pathogens that cause diseases and because authentically "synthetic" vaccines obtained using a variety of molecular antigens only constitute a subgroup of vaccines that have more recently been developed (Jones 2015). However, we can still say that, in the case of vaccination, naturalness bias manifests itself in the tendency to see natural remedies or even the natural germs themselves (i.e., germs that naturally infect people) as preferable to vaccines, which consist of the same germs (either live or inactivated) but are produced in "synthetic" laboratory conditions. DiBonaventura and Chapman showed that naturalness bias, as revealed by people's preference for a herbal drug over a chemically identical synthetic drug, was negatively correlated with participants' intention to obtain a flu vaccine. In the same way, they showed that omission bias, as revealed by parents' refusal of vaccines carrying a risk of iatrogenic disease lower than the risks entailed by the possibility of catching the disease without vaccination, was negatively correlated with the intention to vaccinate. One study found that "[t]he association between non-vaccination and omission bias is not peculiar to those with more or less education, although the more educated respondents (...) were more likely to resist vaccination" (Asch et al. 1994, p. 121). While it is true that correlation (between biases and vaccination decisions) is not the same as causation, it is reasonable to suppose that these biases do play a role in determining vaccination decisions and that therefore such decisions are not rational or based on knowledge about vaccines. This seems to be confirmed, at least with regard to omission bias, by another study that analysed omission bias in vaccination decisions by observing how it affects parents' sense of responsibility for the health outcomes of their children. The study (Ritov and Baron 1990) found that many parents would feel more responsible for the hypothetical death of their child if the death were caused by a vaccine they decided to administer to the child than if the child's death were caused by the very disease against which they decided not to vaccinate. The fact that the same outcome, resulting in both cases from their decision, is associated with a different sense of responsibility depending on whether it is the result of an action or an omission seems to suggest that there is an omission bias at play here. In the qualitative part of the study, a subject said: "I feel that if I vaccinated my kid and he died I would be more responsible for his death than if I hadn't vaccinated him and he died—sounds strange, I know. So I would not be willing to take as high a risk with the vaccine as I would with the flu" (Ritov and Baron 1990, p. 275).

It is not unreasonable, then, to suppose that at least part of the opposition to vaccines is explained not so much by the standard reasons offered by people in surveys about motivations for vaccine refusal or vaccine delay, but by some irrational or biased stance. In other words, concerns about vaccines' safety or effectiveness are likely to be post hoc rationalizations of irrational stances. Granted, it might be argued that a preference for bad outcomes resulting from omission over bad outcomes resulting from action or a preference for the natural over the non-natural (whatever this is taken to mean) do not constitute "biases" as I have defined the concept here. After all, these preferences might be the result of careful ethical reflection rather than of an unanalysed prejudice—which of course does not rule out that the reflection be mistaken; the point is simply that a decision can be irrational and/or unethical without necessarily being the product of some bias. I do not know in what proportion people who refuse vaccination are biased and in what proportion instead they have a reasoned preference for omission over action and for the natural over the unnatural. What I want to highlight is simply that these types of preferences based on allegedly morally relevant distinctions (act/omission; natural/unnatural) are typically not mentioned when people are surveyed about the reasons why they refuse vaccination. This fact seems to suggest, at the very least, that the reasons people offer for their refusal of vaccines do not fully explain their choices and that therefore there is at least an irrational element in such choices not to vaccinate themselves or their children.

Herd Immunity as a Public Good

According to many advocates of coercive vaccination policies, the ultimate goal of such policies should be herd immunity. More precisely, consistently with a principle of "least restrictive alternative", these authors think that states should implement the least coercive policy that is necessary to achieve herd immunity, even if the least restrictive policy entails some level of coercion (e.g., Flanigan 2014; Navin 2015; Pierik 2016). In Chap. 3, I will examine what the principle of "least restrictive alternative" implies with regard to which vaccination policies should be prioritized in the attempt to realize herd immunity from any infectious disease. In Chap. 4, I will question the assumption that vaccination policies should aim *only* at herd immunity. But in order to properly assess the importance of herd immunity and how herd immunity gives people the opportunity to free-ride, thus creating a collective action problem that needs to be regulated through specific—and, if necessary, coercive—policies, it is useful to take a closer look at what herd immunity is and analyse its nature of public good.

Herd immunity is, quite simply, a form of indirect protection from infectious disease. Herd immunity is obtained when a large enough portion of the population is vaccinated, preventing germs from circulating and thereby rendering an infectious disease very unlikely to spread (Fine et al. 2011; Kim et al. 2011). The vaccination coverage rate required for herd immunity varies for different diseases; for example, for measles it ranges between 90% and 95% and for polio between 80% and 85%.

Interestingly, a survey (Sobo 2016) conducted among parents in some US states found that although most parents (70%) were familiar with the notion of "herd immunity", most of these parents did not think it was a reliable measure of safety from infectious disease. In a sense, there is an element of truth in this belief: herd immunity does not offer the same level of individual protection as individual vaccination does and hence is not an equivalent alternative to vaccination. However, herd immunity remains the best form of protection for certain individuals who cannot be vaccinated for medical reasons; for example, the case of the Italian high school class vaccinated against the flu to protect Simone is a case of herd immunity realized on a small scale in order to protect a vulnerable individual.

Now, there are practical problems with relying on herd immunity as a measure for protecting public health and vulnerable individuals. Most notably, the more the rate of international travels intensifies, the less meaningful and useful herd immunity becomes as a preventive measure. With people travelling and moving from one region, state, or continent to the other at an unprecedented rate, it becomes increasingly difficult to identify the relevant community within which herd immunity should be achieved: in one sense, the world has become one big community in a way in which it was not until relatively recently. Simone was protected against the flu only as long as he stayed within his classroom and as long as no out-group unvaccinated individual entered the classroom. If this scenario seems unrealistic when we think of a school class, it is also unrealistic in the large-scale scenario of our globalized world. Ideally, herd immunity would need to be achieved at the global level and not just within national boundaries. However, since vaccination policies are typically implemented at the national level, as things stand now, the only way to ensure that vulnerable individuals are protected as much as possible in the globalized world is that each nation realizes herd immunity within its jurisdiction.

It is important to understand the concept of "herd immunity" not only from a medical and scientific point of view but also with regard to its social and ethical relevance. In Chap. 2, I will explain how, given certain ethical premises, the existence or prospect of herd immunity grounds an individual moral obligation to be vaccinated or to vaccinate one's children. For the moment, in order to prepare the ground for such discussion, it will be useful to say something more about the ethical and social significance of herd immunity and what it means for herd immunity to have "ethical" and "social" significance.

In order to do this, we need first to reflect on its nature of collective good and of public good (Dawson 2007). That herd immunity is a collective good means, quite simply, that the cooperation of a sufficiently large number of people is required to realize it (Dawson 2007, pp. 167–168): no individual or small group of individuals can realize herd immunity. That herd immunity is a public good means that it is both *non-excludable* and *non-rivalrous*. These are technical terms borrowed from the field of economics. Simply put, a good is non-excludable if no one can easily be prevented from benefitting from it (it is often possible to prevent individuals from benefitting from public goods, but when this would be difficult or very costly, the good is considered non-excludable); and a good is non-rivalrous if any individual benefitting from it does not diminish the extent to which other individuals benefit as well. A firework show is an example of a public good. However, firework shows are not important public goods because they do not significantly impact on the well-being of those who enjoy them, and certainly they are not necessary in order to fulfil some fundamental right of individuals; therefore, we cannot say that society or institutions have a moral obligation to provide firework shows. Important public goods are instead things like clean air, national defence, and flood defences; these are the public goods that, for the sake of everyone's interest, a society ought to maintain through a joint effort of its members and/or through institutional interventions. Herd immunity from infectious diseases belongs to this category of important public goods. In Chap. 2, we will see how herd immunity gives rise to collective, individual, and institutional obligations.

Earlier, I said that herd immunity has both social and ethical relevance. It is easy to see in what sense herd immunity has *social* relevance: society as a whole is affected by whether or not herd immunity from any infectious disease exists. A well-functioning society requires a certain level of public health. Herd immunity produces benefits at the societal level because it improves public health and reduces the public costs of healthcare as well as the economic losses associated with illnesses. Everybody benefits from living in a society with herd immunity and therefore with a low rate of infections, regardless of whether they are vaccinated. More precisely, there are three ways in which herd immunity benefits individuals and society. First and foremost, herd immunity protects the unvaccinated. Second, and perhaps less obviously, herd immunity protects the vaccinated as well, since no vaccine is 100% effective; for example, for the 2018 flu season, the estimate of vaccine effectiveness against influenza A (H3N2) was only 10% (Paules et al. 2018), and the pertussis vaccine is only 70% effective during the first year and its effectiveness decreases to 30–40% after four years (CDC 2017). Third, everybody benefits from herd immunity because living in a society with herd immunity means that less public resources need to be diverted to treat sick people; for example, in the US, the flu costs annually US$10.4 billion for hospitalizations and outpatient visits, and the total economic cost associated with annual influenza epidemics, including loss of earning caused by illness, has been estimated to be US$87.1 billion (Molinari et al. 2007). Preserving or realizing herd immunity is therefore important for society, and there are strong ethical as well as economic reasons for a collective to realize herd immunity.

Meanwhile, the *ethical* relevance of herd immunity is explained by its nature of public good as well as by its being a matter of collective, rather than individual responsibility. I will discuss the former aspect here, and the latter in the next chapter. Like all public goods, herd immunity gives rise to a *free-riding problem*. This problem arises when someone would benefit from a certain good regardless of whether they contribute to the good. In such circumstances, a person does not have any incentive to make their contribution; instead, they have an incentive to "take a free ride". The free-riding problem, in turn, gives rise to a collective action problem, that is, a problem that arises because too many people do or fail to engage in a certain action: it is rational for anyone not to contribute to a public good, but too many people acting rationally and failing to contribute compromise the very same public good. The problem arises in the case of vaccination precisely because there is no incentive, and indeed it might be irrational (at least in terms of cost-benefit analysis) for any person to contribute to herd immunity through vaccination when they know herd immunity already exists, since they would be (sufficiently) protected from infectious disease anyway. This mismatch between individual interest and collective interest is precisely where the ethical relevance of herd immunity lies: *if* the preservation or the realization of herd immunity posed any requirement on people at all, it would require (at least some) people to make their contribution to the public good regardless of whether vaccination would be (significantly) beneficial to them or of whether the risk/benefit assessment of vaccination is favourable. Therefore, being vaccinated is often primarily an *ethical* choice: its social importance requires individuals to make a choice for the sake of the public good, rather than exclusively for the sake of their own individual benefit. Besides, because individuals do not have strong enough incentives to contribute to public goods, and because we cannot expect that a large enough number of individuals behave ethically and make their selfless contribution to public goods—free-riding is often simply too tempting—typically the protection or realization of public goods requires institutions to enforce specific policies that, if necessary, coerce individuals into making their contribution. In Chaps. 3 and 4 I will discuss the ethical justifiability of different possible vaccination policies.

Of course, as said above, one might observe here that individuals do stand to benefit from vaccination, because vaccination confers them protection (though not 100% protection) against infectious diseases, and therefore the benefit is primarily individual, and therefore vaccination is rational from the point of view of individual interest; only secondarily, and as a side effect, vaccination contributes to benefitting others. However, there are two considerations to be made here: first, many individuals do not think that they (or their children) would benefit from vaccination, so to them, vaccinating would still be seen as something that goes against their personal interest, and second, as I have mentioned earlier and as we will see better in Chap. 2, vaccination ceases to be individually overall beneficial when vaccination coverage rates are sufficiently high and the small risks of vaccination outweigh the risk of catching the disease and the risks associated with the disease (which oftentimes include the risk of death).

But as mentioned above, the concept of herd immunity is also ethically relevant because realisation or preservation of herd immunity is a matter of collective, rather than individual responsibility: on a large population, no single individual can, by herself, make a significant difference to whether herd immunity exists. How can individuals have an ethical obligation to make an insignificant contribution? So far, I have only said that *if* individuals have a reason to contribute to herd immunity, this has to be an ethical reason, that is, a reason not based (exclusively) on self-interest. But I have not yet demonstrated that individuals *do* have such a reason or ethical obligation. Actually, at a first glance, there seem to be no good reason or ethical obligations to contribute, regardless of whether one has the selfish desire to free-ride: one more vaccinated individual would not make a significant difference to whether a certain community realizes herd immunity or not. What is the ethical reason for being vaccinated or for vaccinating one's children, then? This is the question I will address in the next chapter.

REFERENCES

Asch, D. A., Baron, J., Hershey, J. C., Kunreuther, H., Meszaros, J., Ritov, I., & Spranca, M. (1994). Omission Bias and Pertussis Vaccination. *Medical Decision Making: An International Journal of the Society for Medical Decision Making, 14*(2), 118–123.

Bambery, B., Douglas, T., Selgelid, M. J., Maslen, H., Giubilini, A., Pollard, A. J., & Savulescu, J. (2017). Influenza Vaccination Strategies Should Target Children. *Public Health Ethics, 11*, 221–234.

Carson, P. J., & Flood, A. T. (2017). Catholic Social Teaching and the Duty to Vaccinate. *The American Journal of Bioethics: AJOB, 17*(4), 36–43.

CDC (Centers for Disease Control and Prevention). (2018). *Multiple Vaccines and the Immune System.* Retrieved April 2018, from https://www.cdc.gov/vaccinesafety/concerns/multiple-vaccines-immunity.html

CDC (Centers for Disease Control and Prevention). (2017). Pertussis FAQ. Retrieved April 2018, from https://www.cdc.gov/pertussis/about/faqs.html

Clarke, S., Giubilini, A., & Walker, M. J. (2017). Conscientious Objection to Vaccination. *Bioethics, 31*(3), 155–161.

Dawson, A. (2007). Herd Protection as a Public Good: Vaccination and Our Obligations to Others. In A. Dawson & M. Verweij (Eds.), *Ethics, Prevention, and Public Health* (pp. 160–178). Oxford: Clarendon Press.

Dawson, A. (2011). Vaccination Ethics. In A. Dawson (Ed.), *Public Health Ethics. Key Concepts and Issues in Policy and Practice* (pp. 143–153). New York: Cambridge University Press.

Di Bonaventura, M., & Chapman, G. B. (2008). Do Decision Biases Predict Bad Decisions? Omission Bias, Naturalness Bias, and Influenza Vaccination. *Medical Decision Making, 28*(4), 532–539.

Dubé, E., Laberge, C., Guay, M., Bramadat, P., Roy, R., Bettinger, J. A. (2014a). Vaccine Hesitancy. *Human Vaccines & Immunotherapeutics 9*(8), 1763–1773.

Dubé, E., Gagnon, D., Nickels, E., Jeram, S., & Schuster, M. (2014b). Mapping Vaccine Hesitancy—Country-Specific Characteristics of a Global Phenomenon. *Vaccine, 32*(49), 6649–6654.

Durrheim, D., & Crowcroft, N. (2017). The Price of Delaying Measles Eradication. *The Lancet Public Health, 2*(3), e130–e131.

ECDC. (2018, January). *Measles in the EU/EEA: Current Outbreaks, Latest Data and Trends.* Retrieved May 2018, from https://ecdc.europa.eu/en/news-events/measles-eueea-current-outbreaks-latest-data-and-trends-january-2018

Favin, M., Steinglass, R., Fields, R., Banerjee, K., & Sawhney, M. (2012). Why Children Are Not Vaccinated: A Review of the Grey Literature. *International Health, 4*(4), 229–238.

Fine, P., Eames, K., & Heymann, D. (2011). "Herd Immunity": A Rough Guide. *Clinical Infectious Diseases, 52*(7), 911–916.

Flanigan, J. (2014). A Defense of Compulsory Vaccination. *HEC Forum, 26*, 5–25.

Giubilini, A., Douglas, T., & Savulescu, J. (2017). Liberty, Fairness, and the 'Contribution Model' for Non-medical Vaccine Exemption Policies: A Reply to Navin and Largent. *Public Health Ethics, 10*(3), 235–240.

Giubilini, A., Douglas, T., & Savulescu, J. (2018). The Moral Obligation to Be Vaccinated: Utilitarianism, Contractualism, and Collective Easy Rescue. *Medicine, Health Care, and Philosophy.* https://doi.org/10.1007/s11019-018-9829-y.

Greene, J. (2013). *Moral Tribes. Emotion, Reason, and the Gap Between Us and Them.* London: Atlantic Books.

Harmsen, I. A., Mollema, L., Ruiter, R. A. C., Paulussen, T. G. W., de Melker, H. E., & Kok, G. (2013). Why Parents Refuse Childhood Vaccination: A Qualitative Study Using Online Focus Groups. *BMC Public Health, 13*, 1183.

Hough-Telford, C., Kimberlin, D. W., Aban, I., Hitchcock, W. P., Almquist, J., Kratz, R., & O'Connor, K. G. (2016). Vaccine Delays, Refusals, and Patient Dismissals: A Survey of Pediatricians. *Pediatrics, 138*(3), 2016–2127.

Jones, L. (2015). Recent Advances in the Molecular Design of Synthetic Vaccines. *Nature Chemistry, 7*, 952–960.

Kahan, D. (2014). *Vaccine Risk Perceptions and Ad Hoc Risk Communication: An Empirical Assessment* (CCP Risk Perception Studies Report No. 17, Yale Law & Economics Research Paper # 491). SSRN: https://ssrn.com/abstract=2386034

Kim, T. H., Johnstone, J., & Loeb, M. (2011). Vaccine Herd Effect. *Scandinavian Journal of Infectious Diseases, 43*(9), 683–689.

Largent, M. (2012). *Vaccine. The Debate in Modern America.* Baltimore: Johns Hopkins University Press.

Meszaros, J. R., Asch, D. A., Baron, J., Hershey, J. C., Kunreuther, H., & Schwartz-Buzaglo, J. (1996). Cognitive Processes and the Decisions of Some Parents to Forego Pertussis Vaccination for Their Children. *Journal of Clinical Epidemiology, 49*(6), 697–703.

Molinari, N.-A. M., Ortega-Sanchez, I. R., Messonnier, M. L., Thompson, W. W., Wortley, P. M., Weintraub, E., & Bridges, C. B. (2007). The Annual Impact of Seasonal Influenza in the US: Measuring Disease Burden and Costs. *Vaccine, 25*(27), 5086–5096.

Navin, M. (2015). *Values and Vaccine Refusal: Hard Questions in Ethics, Epistemology, and Health Care.* New York: Routledge.

Navin, M., & Largent, M. (2017). Improving Nonmedical Vaccine Exemption Policies: Three Case Studies. *Public Health Ethics, 10*(3), 225–234.

NHS (National Health Service UK). (2016). *Vaccinations.* Retrieved May 2018, from https://www.nhs.uk/conditions/vaccinations/

Omer, S. B., Salmon, D. A., Orenstein, W. A., deHart, M. P., & Halsey, N. (2009). Vaccine Refusal, Mandatory Immunization, and the Risks of Vaccine-Preventable Diseases. *The New England Journal of Medicine, 360*(19), 1981–1988.

OVG (Oxford Vaccine Group). (2018). *Inactivated Flu Vaccine.* Retrieved February 26, 2018, from http://vk.ovg.ox.ac.uk/inactivated-flu-vaccine

Paules, C. I., Sullivan, S. G., Subbarao, K., & Fauci, A. S. (2018). Chasing Seasonal Influenza—The Need for a Universal Influenza Vaccine. *The New England Journal of Medicine, 378*(1), 7–9.

Perry, R. T., Murray, J. S., Gacic-Dobo, M., Dabbagh, A., Mulders, M. N., Strebel, P. M., et al. (2015). Progress Toward Regional Measles Elimination—Worldwide, 2000–2014. *MMWR. Morbidity and Mortality Weekly Report, 64*(44), 1246–1251.

Pierik, R. (2016). Mandatory Vaccination: An Unqualified Defense. *Journal of Applied Philosophy.* https://doi.org/10.1111/japp.12215.

Ritov, I., & Baron, J. (1990). Reluctance to Vaccinate. Omission Bias and Ambiguity. *Journal of Behavioural Decision Making, 3*, 263–277.

Salmon, D. A., Moulton, L. H., Omer, S. B., DeHart, M. P., Stokley, S., & Halsey, N. A. (2005). Factors Associated with Refusal of Childhood Vaccines Among Parents of School-Aged Children: A Case-Control Study. *Archives of Pediatrics & Adolescent Medicine, 159*(5), 470–476.

Seither, R., Calhoun, K., Street, E. J., Mellerson, J., Knighton, C. L., Tippins, A., & Underwood, J. M. (2017). Vaccination Coverage for Selected Vaccines, Exemption Rates, and Provisional Enrollment Among Children in Kindergarten—United States, 2016-17 School Year. *MMWR. Morbidity and Mortality Weekly Report, 66*(40), 1073–1080.

Smith, P. J., Humiston, S. G., Marcuse, E. K., Zhao, Z., Dorell, C. G., Howes, C., & Hibbs, B. (2011). Parental Delay or Refusal of Vaccine Doses, Childhood

Vaccination Coverage at 24 Months of Age, and the Health Belief Model. *Public Health Reports, 126*(Suppl 2), 135–146.

Sobo, E. (2016). What Is Herd Immunity, and How Does It Relate to Pediatric Vaccination Uptake? US Parent Perspectives. *Social Science and Medicine, 165*, 187–195.

Vanderslott, S. (2017a, April 24). Despite Scepticism, Europe Has High Vaccination Rates—But It Shouldn't Be Complacent. *The Conversation.* Retrieved March 19, 2018, from https://theconversation.com/despite-scepticism-europe-has-high-vaccination-rates-but-it-shouldnt-be-complacent-75169

Vanderslott, S. (2017b). Anti-vaxxer Effect on Vaccination Rates Is Exaggerated. *The Conversation.* Retrieved March 19, 2018, from https://theconversation.com/anti-vaxxer-effect-on-vaccination-rates-is-exaggerated-92630

Wang, E., Clymer, J., Davis-Hayes, C., & Buttenheim, A. (2014). Nonmedical Exemptions from School Immunization Requirements: A Systematic Review. *American Journal of Public Health, 104*(11), e62–e84.

Wenger, O. K., McManus, M. D., Bower, J. R., & Langkamp, D. L. (2011). Underimmunization in Ohio's Amish: Parental Fears Are a Greater Obstacle than Access to Care. *Pediatrics, 128*(1), 79–85.

WHO. (2014). *Report of the SAGE Working Group on Vaccine Hesitancy.* Retrieved March 2018, from http://www.who.int/immunization/sage/meetings/2014/october/1_Report_WORKING_GROUP_vaccine_hesitancy_final.pdf

WHO. (2018). *Europe Observes a 4-fold Increase in Measles Cases in 2017 Compared to Previous Year.* Retrieved March 15, 2018, from http://www.euro.who.int/en/media-centre/sections/press-releases/2018/europe-observes-a-4-fold-increase-in-measles-cases-in-2017-compared-to-previous-year

Vaccination and Herd Immunity: Individual, Collective, and Institutional Responsibilities

Abstract This chapter discusses the relation between collective, individual, and institutional responsibilities with regard to the realization of herd immunity from certain infectious diseases. The argument is put forth that there is a form of collective moral obligation to realize herd immunity, that there is a principle of fairness in the distribution of the burdens of collective obligations, and that such principle entails that each of us has the individual moral responsibility to make their fair contribution to herd immunity through vaccination. These individual moral obligations, in turn, entail a further individual obligation to support policies aimed at realizing herd immunity. The chapter concludes with a suggestion that the individual moral obligations to support such policies generate an institutional responsibility to implement them.

Keywords Vaccination • Herd immunity • Collective responsibility • Individual responsibility • Institutional responsibility

© The Author(s) 2019
A. Giubilini, *The Ethics of Vaccination*, Palgrave Studies
in Ethics and Public Policy,
https://doi.org/10.1007/978-3-030-02068-2_2

ESTABLISHING ETHICAL RESPONSIBILITIES IN THE CONTEXT OF VACCINATION

This chapter is about the ethical obligations or responsibilities[1] pertaining to vaccination of three different actors. As we shall see in more detail later, there are three possible bearers of ethical obligations: individuals, collectives, and institutions (such as states). Thus, one could ask whether any individual has a moral obligation to have themselves or their children vaccinated against common infectious diseases, such as the seasonal flu, mumps, measles, rubella, and any communicable diseases that pose major threats to the health and even survival of individuals. Alternatively, one could ask whether our community, as a collective, has a *collective* ethical obligation to realize herd immunity and what it means to have a collective obligation. Or, again, one could ask whether institutions have an ethical obligation to enforce policies that ensure a community's realization of herd immunity against certain diseases. Since I am looking for a philosophical justification for any such ethical obligation, the central parts of this chapter will be rather technical in philosophical terms.

Unsurprisingly, the fact that vaccination decisions need to be taken at three different levels—individual, collective, and institutional—generates conflicts of values within and between these levels. In particular, when grounding any ethical obligation to be vaccinated or to vaccinate one's children, and when legitimizing any coercive policy that forces individuals to vaccinate themselves or their children, a first ethical problem may arise from a conflict between individual best interest and individual autonomy. A second ethical problem may arise from a conflict between individual autonomy and public health. If my arguments are sound, I will, by the end of the chapter, have solved both problems and provided a philosophical justification for certain ethical obligations at each of the three levels. Before presenting this philosophical justification, however, let me say something more about the two ethical problems I have just mentioned, so as to give the reader a clearer view of the challenges ahead.

Let me start with the first problem, namely, the possible conflict between autonomy and best interest. With regard to child vaccination, one might think that it is quite uncontroversial to say that there is an individual ethical obligation to vaccinate one's children in order to protect their health: after all, being protected against an infectious disease seems

---

[1] From now on, I will use the terms "moral" and "ethical", as well as the terms "obligation" and "responsibility" (when referring to forward-looking responsibility), interchangeably.

to be in a child's best interest, and parents have an ethical obligation to act in the best interest of their children, at least when doing so requires reasonable efforts and the interest being protected is deemed important enough (as seems to be the case for an interest in health preservation and as is certainly the case for an interest in survival). Unfortunately, it is more controversial than it might appear to argue that the best interest of a child can ground an ethical obligation to vaccinate one's children. For one thing, as we shall see in a moment, in certain circumstances it is not so clear that it is in a child's best interest to be vaccinated. Furthermore, even if vaccination is in a child's best interest, some parents might still claim that they have the right to make autonomous choices about their children's health and about what goes into their children's body. What I have just said about child vaccination applies *a fortiori* to adult vaccination. While in the case of child vaccination one might argue that parental refusal to vaccinate their children presents a conflict between a child's best interest and parents' right to make autonomous choices about their children's health, the same refusal in the case of adult vaccination does not seem as ethically problematic. At least according to contemporary liberal ethics, principles of self-determination and of bodily integrity outweighs any paternalistic consideration about a competent adult's best interest: if a competent adult autonomously decides not to be vaccinated, the fact that vaccination might be in her best interest does not seem to imply that the adult in question ought to be vaccinated or ought to be forced to be vaccinated. Besides, as was the case with child vaccination, it is not so clear that adult vaccination is always in the adult's best interest.

Since this last claim has probably raised an eyebrow or two, allow me to elucidate it. Of course, vaccine denialists do share the view that parents have a moral obligation to protect their children's health; however, as we have seen in the first chapter, they do not believe that vaccination sufficiently protects their children's health. While vaccine denialists undoubtedly overestimate the risks of vaccination, we have to concede that, as a matter of fact and for reasons different from the ones they defend, sometimes it is really not in a child's best interest to be vaccinated. How so? Even those of us who see vaccines favourably cannot deny the fact that vaccines can have side effects, some of which can occasionally be quite severe. Vaccine injury compensation funds (Mello 2008) adopted in many countries, such as the US and the UK (Looker and Kelly 2011) exist precisely because side effects can happen. For example, the MMR vaccine can cause anaphylactic reactions, and according to some controversial evidence (CDC 2018a). The seasonal flu vaccine can cause Guillain-

Barre Syndrome (GBS), a serious autoimmune disorder that in its most serious forms can cause paralysis. Now, these possible side effects do not by themselves imply that vaccines are not safe and that they are not in children's best interest, given that their probability is extremely low and that it needs to be weighed against the probability of experiencing the severe, and sometimes lethal, consequences of infectious diseases. Anaphylactic reactions occur in less than one in a million individuals vaccinated with the MMR, but we need to consider that measles "can lead to serious complications such as pneumonia and encephalitis (inflammation of the brain). In addition, measles infection damages and suppresses the whole immune system", and "in high income regions of the world such as Western Europe, measles causes death in at least 1 in 5000 cases, but as many as 1 in 100 will die in the poorest regions of the world" (Oxford Vaccine Group 2016). Overall, then, there is wide consensus in the scientific community that "[g]etting MMR vaccine is much safer than getting measles, mumps or rubella" (CDC 2018a). The same goes for the seasonal flu vaccine. The risk of GBS is only one or two cases per million people vaccinated, but GBS can also occur, and actually is more common (though still very rare), after flu infections (CDC 2018b). Moreover, influenza kills between 290,000 and 650,000 people each year (WHO 2018). These considerations suggest that in spite of the risks, vaccination would often be safer than non-vaccination and would thus be in an individual's best interest. If there is a risk of measles pandemic, it is safer and rational for an individual to be vaccinated or to vaccinate her children against measles. However, there is a threshold of vaccination coverage rate in one's community after which the trade-off between risks of vaccination and risks of the vaccine-preventable infectious disease no longer favours vaccination. When a sufficiently large portion of the population (say, 99.99%) is vaccinated, the risk of being infected becomes so low that it is outweighed by the very low risks associated with vaccination. More generally, the higher the proportion of vaccinated individuals in a given population, the lower the payoff for taking the risks associated with vaccination; and after a certain threshold, the risks associated with vaccination will necessarily be lower than the risks associated with the disease. In such circumstances, parents' moral obligation to act in the best interest of their children entails a moral obligation *not* to vaccinate them. Therefore, if the number of vaccine denialists is sufficiently small, they would be right—albeit for the wrong reasons—to claim that vaccination is not in their children's best interest.

Let us move to the second ethical problem mentioned above, namely, the conflict between individual autonomy and public health. One could argue that the right to make autonomous decisions, including autonomous decisions over one's body, is limited by a harm principle: my liberty is not absolute, but is instead constrained by other people's equal liberty and other people's prima facie right not to be harmed by my behaviour. Indeed, it would be hard to find a reasonable ethical or political theory that does not endorse or at least is not consistent with the harm principle so formulated. At a first glance, the principle seems to apply to the case of vaccination, too: I do not have a right to autonomously decide not to vaccinate myself or my children because vaccine refusal could harm other people by exposing them to preventable infectious diseases. However, the situation might be more problematic than it might initially appear. For one thing, some people might object to the idea that there is a duty to protect other members of the community against diseases. They might believe that it would be a good thing to do so, but not that there is a moral obligation to do so, let alone that there should be a legal obligation or some other form of state coercion; for example, they might turn the argument based on the harm principle upside down and argue that what would be required of me in order to protect others would violate some of my fundamental rights (such as a right to bodily autonomy or bodily integrity). Furthermore—and more interestingly from a philosophical point of view—even assuming that there is an individual moral obligation to protect others against infectious diseases, some might deny that this obligation grounds an *individual* moral obligation to be vaccinated. The obligation might be collective (the meaning of which I will explain in greater detail later), but not individual. The reason is that one's contribution to herd immunity through individual vaccination is negligible. More specifically, one might argue that, where herd immunity exists, the moral obligation on each individual to be vaccinated would be weak, since the risk of infection for other people would be very small even if a single individual were not vaccinated (Dawson 2007, p. 171). And conversely, some have argued that where vaccination coverage rate is low, there is not much ground for a moral obligation to be vaccinated *within a utilitarian perspective*, considering that the risk that any other individual would be infected is high anyway, even if one decides to vaccinate (Verweij 2005, p. 329). Thus, that the moral obligation to be vaccinated is grounded in considerations of public health and on the harm principle is more controversial than it might initially appear.

Having elucidated the main philosophical and ethical problems that arise when we want to ground ethical obligations with regard to vaccination behaviours, let us now turn to introducing the first two levels at which ethical obligations need to be established: the individual and the collective.

Health, Rights, and Ethical Obligations

All of us have some prima facie rights known as *claim* rights, that is, rights determining prima facie obligations that fall on other people. For example, we have a prima facie claim right not to be harmed by others, which generates a prima facie obligation on others not to harm us. This is an example of a *negative* right. Negative rights are the ones that give rise to *negative* duties, that is, moral obligations to *abstain* from doing something that could harm other people or that could prevent a certain benefit to other people. Positive rights, on the contrary, are the rights that give rise to positive duties, that is, moral obligations to *do* something in order to benefit someone or to prevent harm to someone. Other things being equal, positive duties (and rights) are rarer and more difficult to justify than negative duties (and rights). In other words, other things being equal, it is normally considered morally worse to cause harm by action (thus violating a negative duty) than to let the same type of harm happen by omission (thus violating a positive duty). The justification for the normative relevance of the act/omission distinction is a matter of moral theory and is beyond the scope of this book. If one sticks to the intuition that the act/omission distinction does have some normative relevance, one could argue that since vaccination is a positive action, then from a moral point of view, failing to be vaccinated or to vaccinate one's children is not the same as actively engaging in behaviours that harm or risk harming others. While there is a negative duty to refrain from the latter, some might argue that there is no positive duty to do the former, or at least the duty to do the former is significantly weaker. In particular, one could use the allegedly normative distinction between positive and negative duties to dismiss Jessica Flanigan's famous analogy between foregoing vaccination and randomly shooting a gun in the air (Flanigan 2014). According to Flanigan, the two behaviours are relevantly similar from a moral perspective, in that they both threaten other people's health and life. Therefore, according to her, in the same way as an authority has good reasons for prohibiting random gun firing, it also has equally good reasons for prohibiting non-vaccination. One might appeal to the distinction between positive and negative rights and duties to dismiss this analogy: while randomly firing guns violates a negative duty, failing to be vaccinated or to vaccinate one's children would only represent a failure to act in order to benefit others and might therefore not give rise to any positive duty or positive right. Thus, within this perspective, failing to be vaccinated or to vaccinate one's child cannot be morally equivalent to actively putting other people's health and life at risk.

Is this response based on the act/omission distinction valid? I do not think so. Let us assume for the moment that any single non-vaccination does pose a significant risk to others, comparable to the risk posed by someone randomly firing a gun (as we will see below, this assumption is itself problematic). There are reasons to resist the conclusion that failing to vaccinate a child is, as an inaction, less bad than actively doing something that threatens other people's health and life. More can be said in support of the idea that failing to be vaccinated or to vaccinate one's children does represent a failure to fulfil a stringent moral duty, even if it is only a positive duty. After all, Flanigan might be onto something with her analogy. The fact is that the normative force of the act/omission distinction, even if intuitively valid in most circumstances, does not seem to retain its intuitive and normative force in *all* circumstances. One likely explanation for the intuitive attribution of normative relevance to the act/omission distinction is that, quite obviously, it is often easier to do nothing than to do something. Therefore, when the same types of outcomes are considered, such as the generation of a certain harm or risk of harm to others, the obligation to refrain from doing something that could harm others seems more stringent than the obligation to do something in order to avoid possible harm to others: the easier it is for me to prevent a certain risk of harm, the fewer excuses I have for failing to prevent it. But if the intuitive normative force of the act/omission distinction is indeed explained by the relative higher demandingness of actions over omissions, it follows that the easier and less demanding an action is, the closer a failure to take that action in order to prevent harm to others comes to actively harming others, morally speaking, when the same type of harm is considered. In other words, when the actions required to avoid possible harm to others are sufficiently easy and costless, positive duties become morally equivalent to negative duties. Failing to take action and thus to fulfil a positive duty would be significantly similar, from a moral point of view, to the violation of a negative duty.

The idea I have just formulated is sometimes referred to, more simply and more intuitively, with the term "duty of easy rescue": positive duties of "easy rescue" can be as compelling as negative duties, both ethically and legally (Savulescu 2007). For the moment, let us confine ourselves to the ethical dimension. To recall Peter Singer's famous thought experiment, if I see a child drowning in a pond whom I could easily save at a *comparatively* small cost to me (e.g., at the cost of ruining my new pair of shoes), then I have a moral obligation to save the child (Singer 1972), even if the obligation is an instance of a positive, and not of a negative, duty. The child has a positive right to be saved, considering how *comparatively* easy and costless it is for me to save her. Tim Scanlon made the duty of easy rescue even less demanding by stating that we have an uncontroversial moral duty to do something that involves a slight or moderate sacrifice (in absolute, not in comparative terms) and that can prevent something very bad from happening (Scanlon 1998, p. 224)—a formulation that, unlike the Singerian one, does not entail that we have a moral duty to do something that is very costly to us if the outcome to be prevented is *comparatively* very bad. Here, to make my point as uncontroversial as possible, I will stick with Scanlon's less demanding formulation. What I want to point out is that since vaccination entails a very small cost to individuals and a very large benefit to others in terms of disease prevention, there is a duty of easy rescue to be vaccinated, even if it is a positive duty, in order to protect the categories of vulnerable individuals I have mentioned in the previous chapter. *If* non-vaccination harms or risks harming others, then failing to vaccinate is as bad as positively doing something that harms or risks harming others, as Flanigan's analogy suggests.

But we need to be careful here. As I have noted earlier, the contribution of each individual vaccination to herd immunity is negligible. It is true that if a non-vaccinated individual does infect another individual, then the non-vaccinated individual would be causally and morally responsible for the harm caused to the other (Giubilini et al. 2018) or at least the individual's carers would be. However, as suggested above, where herd immunity does exist, it is very unlikely that a non-vaccinated individual would infect another one; and where vaccination rates are extremely low, a non-vaccinated individual would not make a significant difference to the risk of another individual being infected—if she does not infect a specific individual, someone else probably will and epidemics will occur. The risk that a non-vaccinated individual would actually make a significant difference to the chances that another individual is harmed is therefore significant only if vaccination rates in one's community are within a certain specific range, neither too high nor too low. Only in such cases an individual duty of easy rescue applies and morally requires individuals to be vaccinated. However, if we want to ground an *unconditional* moral obligation for any individual to be vaccinated (except, of course, in the case of medical contraindications), and not just an obligation that is dependent on a contingent risk of harming others, we need to find some other form of moral justification. Does this justification exist? I will argue that it does.

Let us start from the existence of an individual's prima facie claim right not to be infected by a vaccine-preventable disease, when this can be achieved through vaccinations. If an individual cannot be vaccinated, or if a vaccine is not effective in an individual (no vaccine is 100% effective), who is the bearer of the corresponding moral duty not to harm, either by act or omission, the vulnerable individual? Sometimes the obligations corresponding to certain individual rights cannot be fulfilled by individuals, but only by collectives. This is the case with the individual right to be protected from vaccine-preventable infectious diseases: since it is only the public good of herd immunity that can guarantee a sufficiently high level of protection, the obligation in question is the obligation to realize herd immunity. And the realization of herd immunity can only be a matter of collective responsibility (Giubilini et al. 2018). As put by Robert Goodin, "responsibilities get collectivized simply because that is the only realistic way (…) of discharging them" (Goodin 1998, p. 55).

A classic example in the philosophical literature of collectivized responsibilities is Parfit's "Harmless Torturers" case, where each torturer contributes only negligibly to the pain experienced by the victims, but the victims feel pain as a result of the contributions of a sufficiently high number of torturers (Parfit 1984, p. 81). Also in such case, the moral obligation not to inflict pain is collective, and not individual, since by hypothesis each individual torturer is "harmless". But what does it mean to have a collective obligation? Who or what, exactly, is the bearer of this obligation, and what does a collective obligation imply for individual obligations? In the remaining of this chapter, I will attempt to answer such questions.

Aggregate Collective Responsibility and Herd Immunity

In what sense can a collective be responsible or have a collective moral obligation? In particular, I am referring here to the responsibilities of the communities that can realize herd immunity, and therefore of unstructured, loose collections of individuals, rather than the responsibilities of organized, structured groups that can be assimilated to individual agents (List and Pettit 2011). Also, in principle and in an ideal world, the collective in question would have to include the entire global population, because local failures to realize herd immunity could endanger the life or health of the vulnerable people living in a certain area. Therefore, ideally, the type of collective obligation we are looking for would have to be what Bill Wringe called a "global obligation" (Wringe 2014). However, in many areas of the world, and in developing countries especially, access to vaccines can be both very difficult and very expensive. It seems unreasonable to expect that people in poor countries who do not have access to vaccines have a collective moral obligation to contribute to the realization of herd immunity. I will proceed under the assumption that the collective obligation to realize herd immunity, although in principle a "global obligation", given the situation of our world, currently only applies to the group of people with easy access to vaccines.

Now, if it is true that the obligation to realize herd immunity cannot be individual, it also seems problematic to argue that there exists a collective to which such responsibilities can be attributed. Some authors, for example Peter French (1984), have argued that only collectives with a formal decision structure can be the subjects of collective obligations. These types of groups constitute collective entities that, because of their internal structure and decision procedures, count as particular types of *agents* and therefore might bear a form of responsibility (as some have argued, including Pettit 2007, and List and Pettit 2011). However, people who together have the causal power to realize herd immunity constitute simply a random collection of individuals. There is no structured and formal connection or coordination among individuals that render them a collective *agent*. And to the extent that we think that only agents, that is, individuals that can intentionally act, can have the responsibility to act in certain ways, the collective that could realize herd immunity cannot have the moral responsibility to realize herd immunity, at least not in the same sense as agents like a state or a corporation have the responsibility to bring about or prevent certain outcomes. Attribution of collective responsibility to unstructured groups might reflect some form of metaphorical talking, but it is difficult to see how collective responsibility can *literally* be attributed to such groups.

However, according to some, it is not necessary that a group has a structure and an internal organization in order to be considered an agent and therefore a subject of collective obligations. As Sean Aas has suggested, when individuals are prepared to do their part in a collective enterprise "were they to become sufficiently sure that others will do their part as well" (Aas 2015, p. 13), then it makes sense to say that the collective, rather than its individual members, is doing something; hence, the collective amounts to an agent that is, in turn, subject to collective obligations (Aas 2015). However, whether or not this position is philosophically sound, the collective that can realize herd immunity through a large enough number of individual members being vaccinated clearly is not an agent in this sense either: typically, as we have seen in Chap. 1, individuals do not decide whether to be vaccinated or to vaccinate their children on the basis of a belief as to whether others will be vaccinated as well, and therefore they do not form a collective agent in Aas' sense either—although, as we shall see in Chap. 4, the assurance that other people around them are vaccinated removes a psychological barrier to choosing to vaccinate.

According to another view, collective obligations can be characterized as "joint obligations that are jointly owned by [individual] agents together" (Pinkert 2014, p. 189). For example, the obligation to form a circle by definition requires the joint effort of a plurality of subjects, since no individual can by herself form a circle. In this respect, the obligation to realize herd immunity is analogous to the obligation to form a circle, that is, it is a joint obligation. But what kind of agent can be the bearer of joint obligations? According to those who endorse the "joint obligation" understanding of collective responsibility or collective obligations, joint obligations could be attributed only to individuals who can engage in joint actions. By "joint actions", Felix Pinkert means "things that a plurality of agents do together, for example, to form a circle, independent of whether or not they have any specific joint intentions" (Pinkert 2014, p. 191). However, this definition is problematic: forming a circle does seem to require a specific joint intention to form a circle, as a random collection of individuals is unlikely to form a circle as a result of random individual behaviours of its member. More in general, it is difficult to see how something can qualify as a "joint action" without some form of joint or shared intention, that is, the intention to take part in a shared enterprise on the common understanding that everybody else is doing their part as well (this is a very simplified definition of the notion of "shared intention" as presented in a way more sophisticated way by Michael Bratman (1993, p. 106)). However, as I have mentioned in the previous paragraph, realizing herd immunity does not require any joint intention or coordinated action by individual members of a collective. In this respect, realizing herd immunity is different from performing a joint action. Therefore, it is at least problematic to say that the collective obligation to realize herd immunity can be conceptualized as a joint obligation in the sense just presented, if it is true that being subject to joint obligations requires that the individuals of the collective engage in joint actions and if it is true that joint actions, such as forming a circle, require the joint intention and the coordinated efforts of individuals.

And indeed, on Anne Schwenkenbecher's account of joint actions, such actions do presuppose a "joint goal" and "a condition of mutual belief and knowledge regarding other people's contributions to that goal: People who act jointly with others do so because they believe that these others will contribute their share towards the joint goal" (Schwenkenbecher 2013, p. 313). This account closely resembles Aas' aforementioned condition for the existence of collective agents who can be the bearers of collective obligations. But, once again, individuals jointly realizing herd immunity by being individually vaccinated do not typically engage in joint actions in neither of the senses just presented. Indeed, Schwenkenbecher convincingly argues that there is an important difference between cases of individuals capable of engaging in joint actions, which can be subjects to "joint duties" to engage in joint actions, and a type of case analogous to the realization of herd immunity, namely, collectively reducing carbon footprint, which requires that a large number of individuals reduce their individual emission (in the same way as realizing herd immunity requires that a large number of individuals be vaccinated). For one, on Schwenkenbecher's account, "joint action of individuals in groups that are not group agents works best on small to medium scale" (Schwenkenbecher 2013, p. 321), rather than on large scales; thus, even if realizing herd immunity required a joint action, this action would be almost impossible to be carried out, since it would require the cooperation of a very large number of individuals. In addition, and more importantly, mitigating global warming or realizing herd immunity does not require joint actions but only aggregate individual actions. The only situation in which mitigation of global warming or realization of herd immunity would presuppose joint actions, and therefore would be the object of joint duties, would be if the potential contributors to the collective effects were part of a movement where individuals can be thought to share the intention to realize the collective effect and therefore to be acting on the basis of the beliefs that others will contribute as well. We can think—so Schwenkenbecher suggests—of the hypothetical organization that she dubs "Citizens for Climate Change Mitigation" (Schwenkenbecher 2013, p. 315) or, we might propose, a hypothetical "Citizens for the Realization of Herd

Immunity". Since such organizations do not exist in our world—although one might wish that they did, and perhaps the case of the highschool class discussed in Chap. 1 is an example of this—there can be no joint duty to mitigate climate change or to realize herd immunity as things stand now. Thus, the collective obligation to realize herd immunity cannot be conceptualized in terms of a "joint duty" either.

So far, it seems that the obligation to realize herd immunity can be neither genuinely collective, since there is no plausible understanding of collective responsibility that can be attributed to a loose, unstructured collection of individuals; nor individual, since it is not in any individuals' power to realize herd immunity. And yet, if we want to claim that individuals have certain prima facie claim rights (particularly the right to be protected from vaccine-preventable infectious diseases), and if the necessary condition for someone to be protected from infectious disease is that herd immunity from a certain disease is realized, we need to be able to say that there is someone or some entity bearing the corresponding moral obligation.

Some scholars have advanced the idea that when there clearly is a desirable collective outcome but apparently no actual organized or goal-oriented collective entity that could realize it, there is nonetheless a collective agent that might be the bearer of some form of collective moral responsibility. Such a collective agent is merely potential, or putative, rather than actual (May 1998; Isaacs 2011, 2014)—and what this means will be explained in a moment. However, this feature does not exclude that it could be a bearer of putative or potential moral obligations that have the same implications in terms of attribution of moral responsibility to its members as actual collective obligations of actual collective agents do. Let us analyse in some more details, then, the idea that there are putative group agents with the putative collective obligation to realize herd immunity.

A merely putative group agent is formed by the random collection of individuals that *could* turn themselves into an organized group or a goal-oriented group in a way that is obvious and clear to a reasonable person (Held 1970; Isaacs 2011, p. 153), that is, into something that can be considered a collective moral agent, in order to fulfil the putative or potential collective obligation to realize a certain collective outcome as a group agent (Isaacs 2014). Not all random collections are also putative group agents. That the course of action required of the collective must be obvious to the reasonable person (in order for the collective to be a putative group agent with a putative collective obligation) is a necessary qualification in order to avoid the paradoxical implication that *any* random collection of individuals (e.g., the one formed by me, someone living in Nepal, and the father of a friend of mine) can be the subject of putative collective obligations (Isaacs 2011, p. 153). Importantly, putative group agents are distinct from mere random collections because the former, unlike the latter, having the potential to turn themselves into organized groups, can also have the moral duty to form a group agent (Collins 2013). Hence, there are two types of collective moral duties that can be attributed to putative group agents: the duty to turn themselves into an actual organized or goal-oriented group and the duty to perform, as a group, the coordinated action that will realize the outcome that they have an obligation to realize. When a merely putative collective agent fails to turn itself into an actual organized group, that is, into an actual group agent, in order to fulfil a putative collective obligation, it can be held morally blameworthy or retrospectively responsible for its "collective inaction" (May 1990).

An example will help clarify in what sense merely putative group agents can be bearers of putative collective moral obligations. Consider Tracy Isaacs' "coordinated bystander" case (Isaacs 2011, p. 144): in Isaacs' example, four bystanders see six children on a raft hurtling towards a waterfall. They can only save the children through an obvious (to the reasonable person) course of action requiring them to coordinate among themselves; any individual acting in isolation would not be able to save the children. In such cases, it is clear (to the reasonable person) that the individuals ought to act together to save the children by turning themselves into a group that could take action. We can therefore say that the collective is a *putative* group agent with a *putative* collective obligation to organize itself in order to save the children.

According to Isaacs, many of the global challenges we face today, such as global poverty, hunger, and climate change, require collective actions by agents that are merely putative group agents (Isaacs 2014, p. 43). Assuming, for the sake of argument, that this is true, the question is: could we include in the category of putative group agents the collective of individuals who together could prevent or contain the spread of infectious disease by realizing herd immunity? The answer has to be negative: what is required in order to realize herd immunity is that individuals engage in *aggregate* individual actions, rather than in *coordinated* group actions. Only by being aggregately vaccinated can a large enough group of individuals realize herd immunity. Thus, even if Isaacs claims that global problems that seem analogous to the realization of herd immunity, such as climate change, raise putative collective obligations, we can conclude that the notion of putative collective obligation attributable to putative group agents also fails to account for the type of collective obligation entailed by the duty to realize herd immunity.

Thus, we have seen that the characterization of collective obligations as obligations of structured groups, as joint obligations, and as putative collective obligations cannot account for the type of collective obligation that falls on the collective of individuals that can and ought to realize herd immunity. Should we accept the idea that no existing account of collective obligation can be applied to the collective obligation to realize herd immunity and, therefore, that there can be no positive moral duty to realize herd immunity, given that there is no one to whom we can attribute such duty? It seems that the answer would have to be affirmative. However, if so, the only remaining alternative understanding of collective obligation is the one according to which the obligation to realize herd immunity is collective in the merely aggregative sense, that is, in the sense that *each and every* individual member of the collective with the power to realize herd immunity has a moral obligation to contribute to the realization of the collective effect. However, this conception of collective responsibility has problems of its own, because it would imply attributing moral obligations to individuals even when any one of them fulfilling such obligation would not have any significant impact on, and therefore would not substantially contribute to, the collective outcome.

In sum, it seems we are facing an insurmountable conceptual problem, but one with relevant practical implications: neither individuals nor collectives can be attributed moral obligations to protect vulnerable individuals from vaccine-preventable infectious diseases; therefore, no one seems to be under any moral obligation to be vaccinated. However, such problems are not insurmountable. In the next two sections, I will explain, through a metaphysical (section "Aggregate Collective Responsibility and Herd Immunity") and an ethical (section "From Collective to Individual Responsibility: The Metaphysical Arguments") analysis of the relationship between collective and individual obligations, why individuals are under a moral obligation to contribute to herd immunity even when their contribution would be insignificant.

For the moment, I would like to introduce a new label for the peculiar collective character of the moral obligation to realize herd immunity. I will use the expression *aggregative collective obligation* in order to emphasize the "deflationary" sense of "collective" here involved, that is, that the collective in question is *not* to be understood as an independent entity irreducible to the aggregate of its constituent individuals. In other words, collectives have an obligation to realize herd immunity in the sense that the collective obligation is fulfilled through the *aggregate* actions of the collective members. An alternative, though less explicative way of referring to aggregative collective obligation as I understand it here is the one adopted by Gunnar Björnsson (2014). According to Björnsson's terminology, we can say that the collective obligation to realize herd immunity is "essentially shared" by certain individuals. Essentially shared obligations can be considered a particular type of collective obligations attributable to any groups that can and ought to realize certain outcomes through aggregate individual actions, rather than through coordinated group actions. As put by Björnsson, "shared obligations are not necessarily obligations to perform joint actions" (Björnsson 2014, p. 109) and "a shared obligation can be fulfilled without any sense of coordinated or shared agency among the parties" (Björnsson 2014, pp. 109–110). In other words, a shared obligation can be fulfilled when members of a certain collective engage in a certain behaviour in such a way that a certain collective outcome, for example, herd immunity, is realized. Thus, we can say that there is a shared obligation to realize herd immunity or an aggregative collective obligation to realize herd immunity.

The next question I want to address is what this type of collective obligation entails for attribution of individual obligations to be vaccinated (the metaphysical account of the relationship between collective and individual responsibility, presented in the next section), and why the collective obligation to realize herd immunity translates into an individual obligation to be vaccinated to which *all* members of a collective are subject (the ethical account of the relationship between collective and individual responsibility, which I will present in section "From Collective to Individual Responsibility: The Ethical Argument").

From Collective to Individual Responsibility: The Metaphysical Arguments

So far, I have established the existence of a shared or aggregative collective obligation to realize herd immunity. The next question I want to ask is what such collective obligation implies in terms of individual obligations. Most of the arguments about the existence of collective obligations of loose collections and the relationship between collective and individual responsibility can be characterized as metaphysical arguments: they concern the issue as to what makes an obligation "collective" and the nature of the relationship between such collective obligations and the individual obligations of the collective's constituent members. Without the presumption of exhaustiveness, I will present in this section some of the metaphysical conceptions of collective obligation and of its relation with individual obligations. I will apply these considerations to the case of the collective obligation to realize herd immunity and its relation to the individual obligation to be vaccinated.

Let us start with the case of the trapped man, discussed by Virginia Held (1970). In her example, three pedestrians notice a man trapped under a collapsed building. They can save the man by removing the beams that keep him trapped. However, they fail to organize as a collective and to decide which beam to remove first; as a consequence, the trapped man dies. According to Held, the three men are collectively morally responsible for failing to form an organized group that could have saved the trapped man. This example is taken by Held to show that collective responsibility of random collections is simply distributive: each individual is individually responsible for the group's failure. As put by Held, "if random collection *R* is morally responsible for the failure to do *A*, then every member of *R* is morally responsible for the failure to do *A*, although, perhaps, in significantly different proportions" (Held 1970, p. 480). Held applied this principle of distributive collective responsibility to the case of backward-looking responsibility. However, the same principle can be applied to future-looking moral responsibility, that is, moral obligations. Consider, for instance, the analogous "coordinated bystander" case discussed by Tracy Isaacs (2011, 2014), which we have presented above. According to Isaacs, the putative collective obligation has "exactly the same ordering and mediating potential for individual action that an actual collective obligation would" (Isaacs 2011, p. 150). In cases like this, it is clear (to the reasonable person) what the group should do in order to save the children, so the group agent has the putative collective obligation to save the children. In virtue of this putative collective obligation, Isaacs argues, each individual has a moral obligation to *do her part* (Isaacs 2011, p. 151) to contribute to the fulfillment of the putative collective obligation. As Isaacs put it, "this putative collective obligation (…) is a starting point for bridging the apparent gap between seemingly inconsequential individual contributions and new understandings of the part they play in more powerful collective undertakings" (Isaacs 2011, pp. 151–152).

Granted, when we talk of the collective obligation to realize herd immunity, the kind of collective obligation with which we are concerned is, as noted earlier, not a putative collective obligation, but a shared or aggregative collective obligation. However, Isaacs' account of the relationship between the collective and the individual responsibility does not depend on the putative character of the collective obligation in question, but is instead based on the relationship between what can be realized collectively and the obligation of the individuals within the collective to do what is required in order for the collective effect to obtain. Therefore, the same principle bridging collective and individual obligation can be applied to the case of aggregative collective obligations. To use an example which is different from the one introduced by Isaacs but which is analogous to the aggregative type of collective obligation involved in the case of realization of herd immunity, consider an individual's failure to contribute to the prevention of global warming, for example, by avoiding driving just for fun. Such failure makes the individual blameworthy because it is a failure to "do her part in a collective action that could solve global warming" (Isaacs 2011, p. 151). The action here is collective in the same sense in which the action required to realize herd immunity is collective: the "collective action" consists of individual *aggregate* actions. In such cases, according to Isaacs, the failure to make one's contribution to the desirable collective outcome is "not morally excusable because it is mediated by the putative collective obligation to solve global warming" (Isaacs 2011, p. 151); more precisely, to use the terminology we have introduced, it is mediated by the *shared* or the *aggregative* collective obligation to solve global warming. In the same way, we might say that the failure to *contribute* to herd immunity by being vaccinated is not morally excusable because it is mediated by the shared or aggregative collective obligation to realize herd immunity. So it seems that we have established not only a form of shared or aggregative collective moral obligation to realize herd immunity but also an individual obligation to make a contribution to the realization of herd immunity. And as Isaacs explains, "being a possible member of a group that could effectively take action to address an obvious issue that needs addressing can influence a person's individual moral obligations" (Isaacs 2014, p. 57).

Let us consider now a different account of the metaphysical relationship between collective and individual obligations. This is the account put forward by Bill Wringe. According to Wringe, collective obligations of unstructured, loose collectives are explanatorily and ontologically more fundamental than the obligations of individual members of the collective: the former can be used to explain the existence of the latter. As Wringe put it, "it is part of the moral phenomenology that the individual obligations of A and B can be explained by reference to the existence of a collective obligation and by A and B's membership of the relevant collective" (Wringe 2016, p. 485). Moreover, as Wringe argues elsewhere, "it seems plausible that a claim about the obligations of a collective of which I am a member could have a legitimate influence on me in deciding (or perhaps better, could be a reason relevant to deciding) how to respond to a situation which appears to call for a collective action" (Wringe 2010, p. 226). Such relationships between collective and individual obligations can be explained by distinguishing, as Wringe (2016, pp. 224–225) does, between the *subjects* and the *addressees* of collective obligations. The subjects of a collective obligation are those to whom the obligation applies, which might be collectives, such as the collective with the potential for realizing herd immunity. The *addressees* of collective obligations are those whose capacity for deliberation is affected by the existence of the collective obligations, namely, individual members of the collective, such as the individuals who can contribute to herd immunity by being vaccinated. In this view, the individual addressees of a collective obligation "acquire obligations to do things which are appropriately related to the carrying out of the action whose performance would constitute fulfilment of the collective obligation" (Wringe 2010, p. 227). If we extend the same point to shared or aggregative obligations, we could say that, for instance, individual members of collectives with a shared obligation to realize herd immunity, as addressees of the obligation, acquire the individual obligation to do what allow the collective to fulfil the obligation, namely, being vaccinated.

Wringe has formalized the principle connecting collective to individual obligations, that is, the "global supervenience" of collective over individual obligations, as follows:

> If in a particular situation a collective C has an all-out obligation to Phi, then, for any member M of C, and for any set S of possible actions of members of C that, if performed together, would constitute C's Phi-ing, if S includes M's doing A, then M has a pro tanto obligation to do A provided that (a) the other members of C are doing or are reasonably likely to do the actions assigned to them in S or they would be reasonably likely to do these things if M were to do A and (b) M's doing A does not by itself make it less likely that C will Phi. (Wringe 2016, p. 488)

The formulation is rather (perhaps unnecessarily) convoluted, but with a small effort we can see how the principle applies to the cases we are interested in here by replacing "phi" with realizing herd immunity and "A" with being vaccinated. The fact that individual contributions represent the only *means* through which collectives can fulfil their obligations (Wringe 2016, p. 489) suggests that once we have established that there are collective obligations, such obligations generate obligations for individual members to contribute to the collective effect: there is no other way a collective obligation can be fulfilled except through each individual doing their part (Wringe 2014, p. 180). We can call this the "means argument" for the existence of individual obligations to contribute to collective enterprises.

However, the metaphysical accounts of the relationship between collective and individual obligations I have presented here do not address, let alone answer, the question of why a shared or aggregative obligation is supposed to generate an individual obligation that applies to *each and every* individual member of the collective. The question becomes particularly pressing in light of the fact that any individual vaccination is likely to be neither sufficient nor necessary for the fulfilment of the collective obligation to realize herd immunity. To address and answer this type of question, we need an ethical analysis of why a collective obligation generates individual obligations. This will be the subject of the next section.

From Collective to Individual Responsibility: The Ethical Argument

At least in some cases, if everyone contributed to some collective effect, the effect would be over-determined. This makes it difficult to claim that each and every individual in those cases has a moral obligation to contribute. Realization of herd immunity is one such case. It seems therefore possible to question the idea, which I have introduced in the previous section, that collective obligations to realize herd immunity give rise, by their very own nature, to individual obligations to contribute to herd immunity by being vaccinated. As put by Felix Pinkert, individual obligations of the form "you ought to contribute" in the context of collective obligations "imply that you ought to contribute even if not enough others contribute as well, but it is implausible that one ought to perform such pointless actions. In a more sophisticated form, 'you ought to contribute if enough others contribute as well', it turns out that everyone discharges their obligation if no one contributes" (Pinkert 2014, p. 189), which seems absurd.

According to Isaacs, the fact that each individual contribution would make no difference to the prevention of global warming does not rule out that individuals have a moral obligation to contribute to the prevention of global warming (Isaacs 2011, p. 151). The same seems to follow if we apply Wringe's principle connecting collective and individual responsibility to the case of global warming. Presumably, then, Isaacs and Wringe would say the same about individual contributions to herd immunity: in all such cases, they would say that the collective obligation mediates individual obligations. But why would any individual have a moral obligation to make an irrelevant contribution to an important good? Something clearly needs to be added to their account in order to explain how an individual obligation to make an irrelevant contribution can derive from a shared or aggregative collective obligation. In this section I am going to provide what I think is the missing piece of the puzzle, which involves ethical considerations about how the burdens of a collective, or aggregative, or shared obligation ought to be shared.

Indeed, also in Björnsson's account of essentially shared obligations, there is a problematic relationship between the collective (shared) obligations and individual obligations of members of the collective. One example of shared obligation he provides is that of three people who are polluting a lake by using a certain solvent to paint their boats, which is killing the fish in the lake. The fish could be saved only if at least two of them stopped polluting, but not if only one stopped. According to Björnsson, there is in this case a shared obligation to stop polluting, in the sense that the obligation to stop polluting can be fulfilled by individuals behaving in a certain way that does not require shared intention or coordinated actions. Like the case of herd immunity, this is a situation in which the realization of the desirable collective outcome (saving the fish or realizing herd immunity) depends not on what any single individual does, but on what the other members of the collective do: anyone's contribution to the collective effect is insufficient to realize the desirable collective effect. Where one's contribution to the collective outcome is not sufficient for the collective outcome to occur, there is a mismatch "between reasons underlying the shared obligation and individual reasons to contribute to its fulfilment" (Björnsson 2014, p. 108). Thus, the account of shared responsibility endorsed here "makes it intelligible that a group has an obligation even though no individual agent has an obligation to contribute" (Björnsson 2014, p. 118). In other words, relying merely on a metaphysical account of collective obligation and of its relationship with individual obligation leaves us with a situation wherein the collective obligation to realize herd immunity is insufficient to warrant the existence of any individual obligation.

If all this is true, then we need a separate argument for why shared obligations generate individual responsibilities for the members of collectives with the causal power to realize herd immunity. The argument I am going to provide is a fairly simple and straightforward one, based on considerations of fairness in the distribution of the burdens that a shared or aggregative collective obligation entails. Once we assume that there is some kind of collective obligation to bring about a certain desirable outcome, we also have to assume that there arise a certain amount of individual moral obligations that need to be fulfilled in order for the collective or shared obligation to be fulfilled in turn. After all, as mentioned earlier, fulfilling individual obligations to contribute to the collective effect is the only means through which the collective obligation can be fulfilled. Thus, the collective obligation to realize herd immunity generates a certain amount of "burdens": a certain number of individuals will have to be vaccinated. I call vaccination "a burden" in this context because some people are opposed to it and because vaccination does involve some small inconvenience (possible temporary pain of the injection, having to pay a visit to the doctor, potentially a financial cost, minor risk of some side effects, etc.). That said, we need to bear in mind that vaccination also, and indeed primarily, benefits the individual who is vaccinated by giving her immunity from infectious diseases. All in all, vaccination involves very light and certainly bearable individual burdens, which can be vastly outweighed by the individual benefits it entails. In any case, the relevant question, for our purposes, is the question as to how such burdens should be distributed among individuals who form the collective with the moral obligation to realize herd immunity. It is safe to assume that such burdens should be distributed *fairly*, to the extent that we think that fairness is an important value that needs to be taken into account when distributing any kind of burden involved in the realization of important public goods. Thus, fairness demands that *each individual* does whatever she reasonably can in order to contribute to the fulfilment of the collective or shared obligation, regardless of the actual impact any individual action would have on the realization of the collective outcome. In other words, fairness requires that any individual who has the capacity to reasonably bear such burdens makes her fair contribution to the fulfilment of the collective obligation. For instance, in the case of realizing herd immunity, the group of people with the individual obligation to accept a fair share of the burdens will include any individual who does not have any medical condition that would make vaccination supererogatory, or who is not too young or too old to be vaccinated (Giubilini et al. 2018).

This also means that it would be unfair to require those for whom vaccination would be supererogatory to make their contribution to herd immunity. Such request would not be fair because the burden these individuals would have to bear if they were vaccinated would be much greater than the burden borne by other individuals. Individuals who are either immunosuppressed, allergic to vaccination, or too young or old to be vaccinated do not have a fairness-based moral obligation to contribute to the fulfilment of the collective obligation to realize herd immunity. Indeed, they are the very individuals who ought to be protected from the threat of infectious diseases by making sure that enough people around them are vaccinated.

Thus, fairness provides that missing link between aggregative and individual responsibility as discussed earlier: we can say that it is because of a requirement of fairness that a shared or aggregative collective obligation generates individual moral obligations such as the individual moral obligation to be vaccinated. Since considerations of fairness are not primarily about the impact of one's behaviour on others, but about distribution of benefits and costs, they ground an individual moral obligation to be vaccinated even if any individual vaccination would have no significant impact on vaccine coverage rates and on reducing the risk of infection for other people.

One last problem that needs to be addressed when attributing individual responsibilities for vaccination is that often vaccination decisions concern children, not adults, and that it seems problematic to argue that children have fairness-based moral obligations to be vaccinated. Plausibly, one needs to be a competent moral agent in order to be subject to a moral obligation, and children are not competent moral agents at the age at which most vaccinations are typically recommended: simply, they do not have the adequate level of understanding to make informed decisions and to take responsibility. There are exceptions, though: vaccines against meningococcal groups A, C, W, and Y disease are usually recommended for 12-year-old children, who arguably do count as moral agents and are subject to moral obligations. In such cases, the argument for an individual moral obligation to be vaccinated applies directly to such children. What about younger children and infants? Here, the moral obligation in question is not that of being vaccinated, but that of vaccinating one's children. Since young children cannot take responsibility for their actions, it is parents who have to take responsibility on their behalf: parents have a fairness-based obligation to make their fair contribution to herd immunity by vaccinating their children. Not all moral obligations that parents have, as parents, are directed to the best interest of their children. Vaccination is one example: although vaccinating one's children would often promote their best interest, there is a moral obligation to vaccinate one's children that is not grounded in a duty to promote their best interest, but in a duty of fairness towards society. To the extent that parents can, should, and do make decisions on behalf of their children, they also can make moral decisions on behalf of their children, as they in fact often do in many other contexts.

From Individual to Institutional Responsibility

So far, I have argued that the existence of a collective obligation to realize herd immunity, together with a principle of fairness in the distribution of certain burdens, generates an individual moral obligation on each individual member of the collective to make her contribution to herd immunity. But what does it mean, exactly, to make one's contribution to herd immunity? As we have seen, it certainly means to be vaccinated and to vaccinate one's children against the infectious diseases from which individuals have a prima facie claim right to be protected.

But this cannot be the end of the story. In order for the collective obligation to be fulfilled, it is necessary that *enough* individuals be vaccinated and not just that any single individual is vaccinated. In other words, there are individual obligations and there is the collective obligation, but the collective obligation consists in an obligation that a certain minimum number of individuals are vaccinated, and therefore the contribution each individual ought to make is towards this end. Therefore, the question arises as to whether the contribution any individual ought to make should include doing something that makes it the case, or at least makes it more likely, that enough other people are vaccinated as well. If doing so comes at small or reasonable cost to the individual, then it seems that fairness requires that the individual makes this type of contribution, too. By "reasonable", I mean here something that does not involve a too large cost to the individual, consistent with the duty of easy rescue I have discussed previously.

The traditional or common-sense understanding of moral responsibility is an individualistic understanding whereby moral responsibility only has implications for individual behaviour. But this traditional conception does not provide a satisfactory answer to the question of what an individual ought reasonably to do in order to contribute to the fulfilment of a collective obligation. Taking seriously the role of the collective nature of certain responsibilities in shaping individual moral responsibilities requires going beyond this common-sense individualistic account of responsibility, and embracing what we might call a "political" understanding of moral responsibility: individual responsibility, in contexts of collective responsibility, is *a responsibility to do what one reasonably can to ensure that other people also make their contribution* to the desirable collective outcome, for example, to herd immunity.

But what does this mean in practice? How can an individual reasonably contribute to ensuring that the threshold for herd immunity is reached? We can be sure that enough individuals are vaccinated when there are effective vaccination policies in place—and which policies exactly are required depends on what level of state coercion is necessary to realize herd immunity, an issue I will address in the next chapter. Thus, an individual can reasonably contribute to ensuring that enough others do their part by supporting the adequate forms of organization and policies. This means that an individual obligation to make her contribution to a desirable collective outcome entails a prima facie individual obligation to *support policies* that ensure the contribution of a sufficient number of others as well. To support effective vaccination policies means, at the very least, to refrain from hindering the implementation of such policies; thus, for example, protesting against mandatory vaccination, or requesting exemptions from mandatory vaccination, means failing to fulfil one's moral obligation to do what one reasonably can to ensure that herd immunity is realized and that members of one's community are protected from infectious disease. But to support effective vaccination policies also means to urge governments to implement such policies where they are not in place and herd immunity does not exist yet.

Since individuals have a moral duty to support effective vaccination policies, a democratic state has the strongest justification possible for implementing such policies, at least if we accept the rather uncontroversial principle that the legitimization for public policies in democratic states derives from individuals' support. Even where individuals do not actually support vaccination policies, the fact that they ought to support them makes those policies morally legitimate.

In discussing the problem of what should be done to counteract global warming, Walter Sinnott-Armstrong has suggested something resembling the idea I have just put forward. He argued that individuals do not have a moral obligation to avoid taking one's car for leisure drives on Sundays to prevent global warming, because the impact on global warming of any individual driving is negligible; rather, as Sinnott-Armstrong argues, it is governments that should intervene to prevent global warming, if necessary by prohibiting individuals from recreational driving on Sundays. What individuals have a moral obligation to do is simply "to get governments to do their job" (Sinnott-Armstrong 2005, p. 312), that is, as I have said above, to actively support the appropriate policies. Presumably, Sinnott-Armstrong's argument could also be applied to the case of vaccination and herd immunity: individuals' duty to prevent negative collective outcomes includes the duty to support effective policies with the potential for preventing those outcomes. Contrary to Sinnott-Armstrong's thesis, I have provided here an argument, based on fairness, to the effect that individuals also have a moral obligation to contribute to the containment of global warming or to herd immunity by avoiding recreational driving on Sundays or by being vaccinated, respectively; they have these moral obligations even if their individual contribution to the collective cause is negligible. In any case, what matters for the purposes of the present discussion is that individuals fulfil their moral obligations to contribute to desirable collective outcomes (such as herd immunity) also by supporting policies that guarantee that herd immunity is realized.

Now, as we have seen, Sinnott-Armstrong says that individuals ought to "get governments to do their jobs". But what is governments' "job" with regard to vaccination policies? In other words, what are states' institutional responsibilities? Not everybody will agree with the following principle, and not everywhere is this principle equally accepted, but here I will assume that most people will agree with my understanding: a state has the moral responsibility to protect and promote individuals' health, especially that of the most vulnerable people (such as those who cannot be vaccinated), by at least controlling those factors that (1) affect individual health, (2) are not under an individual's control, and (3) that the state can permissibly control. For example, many countries have in recent decades implemented policies prohibiting smoking in public spaces in order to safeguard the health of non-smokers. If I am a non-smoker, other people's smoke in public spaces is a factor that would affect my health, that is not under my control, and that the state can permissibly control. However, in fulfilling its moral duty to protect vulnerable people's health, a state is not morally justified in doing just *anything* in its power; for example, to go back to infectious diseases, a state is normally not morally justified in quarantining individuals with measles or the flu in order to prevent other individuals from being infected. What, then, ought a state to do in order to protect vulnerable people from vaccine infectious diseases by remaining within its ethical boundaries? What are the limits of a state's moral obligation to protect the health of a community? Questions about what a state ought to do are inseparable from questions about what a state may permissibly do in order to fulfil its moral obligation. From what I have said so far, a state may permissibly fulfil its moral responsibility to protect individual and public health by requiring individuals to fulfil their individual moral obligations; for example, since vaccination is an individual moral obligation, as I have argued in this chapter, the state is justified in requesting individuals to be vaccinated in order to realize herd immunity, given that by doing so the state would not be requesting individuals to do anything supererogatory. Besides, state policies aimed at realizing herd immunity are further justified by the fact that individuals have a moral obligation to support such policies. Thus, the argument I have provided suggests that a state has a moral obligation to *at least* ensure that herd immunity is realized within its jurisdiction. Such institutional obligation results from the combination of a moral duty to protect vulnerable individuals' health and the ethical acceptability of vaccination policies that individuals have a moral obligation to support.

CONCLUSION

Before concluding and taking the next step, let me very briefly summarize the content of this chapter. I have argued that (1) individual rights to be protected from vaccine-preventable infectious diseases generate a collective obligation, which I have conceptualized as aggregative or shared responsibility, to realize herd immunity; (2) such collective obligation generates an individual obligation for every member of a community both to be vaccinated, unless there are medical reasons that would make vaccination supererogatory, and to support policies that allow to realize herd immunity; and (3) such individual obligations to support effective vaccination policies, together with the principle that states ought to protect individuals' health at least with regard to those factors that are under its control, generate the institutional responsibility to implement vaccination policies that can at the very least realize herd immunity.

Now, what specific types of policies individuals have a moral obligation to support, and institutions have the responsibility to implement, depends on the efficacy of possible alternative policies in realizing herd immunity and on their moral costs, for example, in terms of liberty infringements and fairness violations. Other things being equal (e.g., if two types of policies are equally effective in realizing herd immunity), less intrusive policies are to be preferred (Verweij ad Dawson 2004) according to a widely shared principle of "least restrictive alternative" in public health. The analysis and application of this principle to vaccination policies will be the topic of the next chapter. The reason why a principle of least restrictive alternative requires a separate discussion is that it raises more problems than it actually solves. In particular, one might ask (1) which vaccination policies can be considered less restrictive than others, and therefore ought to be preferred, and (2) what goal exactly ought to be pursued through vaccination policies, that is, whether herd immunity or something else. I will address these two questions in Chaps. 3 and 4, respectively. In Chap. 3, I will assume the widely shared view that herd immunity should be the goal of vaccination policies, which also follows from the arguments I have provided in this chapter; in Chap. 4, I will suggest that vaccination policies ought to be more ambitious: the fact that in this chapter I have argued that vaccination policies should *at least* aim at herd immunity does not mean that they should not aim at something even more ambitious, if some justification for this more ambitious target can be provided.

References

Aas, S. (2015). Distributing Collective Obligation. *Journal of Ethics and Social Philosophy, 9*, 3.

Björnsson, G. (2014). Essentially Shared Obligations. *Midwest Studies in Philosophy, 38*(1), 103–120.

Bratman, M. E. (1993). Shared Intention. *Ethics, 104*(1), 97–113.

CDC (Centers for Disease Control and Prevention). 2018a. Measles, Mumps, and Rubella (MMR) Vaccine Safety. At https://www.cdc.gov/vaccinesafety/vaccines/mmr-vaccine.html. Accessed 28 Oct 2018.

CDC (Centers for Disease Control and Prevention). (2018b). Guillain Barré Syndrome. At https://www.cdc.gov/vaccinesafety/concerns/guillain-barre-syndrome.html. Accessed 7 Aug 2018.

Collins, S. (2013). Collectives' Duties and Collectivization Duties. *Australasian Journal of Philosophy, 91*(2), 231–248.

Dawson, A. (2007). Herd Protection as a Public Good: Vaccination and Our Obligations to Others. In A. Dawson & M. Verweij (Eds.), *Ethics, Prevention, and Public Health* (pp. 160–178). Oxford: Clarendon Press.

Flanigan, J. (2014). A Defense of Compulsory Vaccination. *HEC Forum: An Interdisciplinary Journal on Hospitals' Ethical and Legal Issues, 26*(1), 5–25.

French, P. (1984). *Collective and Corporate Responsibility*. New York: Columbia University Press.

Giubilini, A., Douglas, T., & Savulescu, J. (2018). The Moral Obligation to Be Vaccinated: Utilitarianism, Contractualism, and Collective Easy Rescue. *Medicine, Health Care, and Philosophy.* https://doi.org/10.1007/s11019-018-9829-y.

Goodin, R. (1998). *Social Welfare and Individual Responsibility: For and Against.* Cambridge: Cambridge University Press.

Held, V. (1970). Can a Random Collection of Individuals Be Morally Responsible? *Journal of Philosophy, 67*(14), 471–481.

Isaacs, T. (2011). *Moral Responsibility in Collective Contexts.* New York: Oxford University Press.

Isaacs, T. (2014). Collective Responsibility and Collective Obligation. *Midwest Studies in Philosophy, 38*(1), 40–57.

List, C., & Pettit, P. (2011). *Group Agency. The Possibility, Design, and Status of Corporate Agents.* Oxford: Oxford University Press.

Looker, C., & Kelly, H. (2011). No-Fault Compensation Following Adverse Events Attributed To Vaccination: A Review of International Programmes. *Bulletin of the World Health Organization, 89*(5), 371–378.

May, L. (1990). Collective Inaction and Shared Responsibility. *Nous, 24*(2), 269–277.

May, L. (1998). Collective Inaction and Responsibility. In P. French (Ed.), *Individual and Collective Responsibility*. Rochester: Schenkman Books.

Mello, M. (2008). Rationalizing Vaccine Injury Compensation. *Bioethics, 22*(1), 32–42.

Oxford Vaccine Group. (2016). *Herd Immunity.* University of Oxford. Retrieved January 18, 2017, from http://www.ovg.ox.ac.uk/herd-immunity

Parfit, D. (1984). *Reasons and Persons.* Oxford: Clarendon Press.

Pettit, P. (2007). Responsibility Incorporated. *Ethics, 117*, 171–177.

Pinkert, F. (2014). What We Together Can (Be Required to) Do. *Midwest Studies of Philosophy, 23*, 187–202.

Savulescu, J. (2007). Future People, Involuntary Medical Treatment in Pregnancy, and the Duty of Easy Rescue. *Utilitas, 19*(1), 1–20.

Scanlon, T. (1998). *What We Owe to Each Other.* Cambridge, MA: Harvard University Press.

Schwenkenbecher, A. (2013). Joint Duties and Global Moral Obligations. *Ratio, 26*, 310–328.

Singer, P. (1972). Famine, Affluence, and Morality. *Philosophy and Public Affairs, 1*(3), 229–243.

Sinnott-Armstrong, W. (2005). It Is Not My Fault: Global Warming and Individual Moral Obligations. In W. Sinnott-Armstrong & R. Howarth (Eds.), *Perspectives on Climate Change* (pp. 221–253). Amsterdam: Elsevier.

Verweij, M. (2005). Obligatory Precautions Against Infections. *Bioethics, 19*(4), 323–335.

Verweij, M., & Dawson, A. (2004). Ethical Principles for Collective Immunization Programs. *Vaccine, 22*, 3122–3126.

WHO (World Health Organization). 2018. Influenza (Seasonal). At http://www.who.int/news-room/fact-sheets/detail/influenza-(seasonal). Accessed 28 Oct 2018.

Wringe, B. (2010). Global Obligations and the Agency Objection. *Ratio, 23*(2), 217–231.

Wringe, B. (2014). From Global Collective Obligations to Institutional Obligations. *Midwest Studies in Philosophy, 38*(1), 171–186.

Wringe, B. (2016). Collective Obligations: Their Existence, Their Explanatory Power, and Their Supervenience on the Obligations of Individuals. *European Journal of Philosophy, 24*(2), 472–497.

CHAPTER 3

Vaccination Policies and the Principle of Least Restrictive Alternative: An Intervention Ladder

Abstract The principle of least restrictive alternative (PLRA) states that policymakers have significant reason to implement the policy that is effective in achieving a certain result and that is least restrictive of individual liberty or autonomy. This chapter provides a ranking of vaccination policies, or an intervention ladder, on the basis of the PLRA, assessing the level of coercion of each type of policy. The ranking of vaccination policies I suggest, in order of increasing restrictiveness or coerciveness, is as follows: persuasion, nudging, financial incentives, disincentives (including withholding of financial benefits, taxation, and mandatory vaccination), and outright compulsion. Each type of policy suggestion is presented with a discussion of the level of restrictiveness or coerciveness involved and the potential effectiveness.

Keywords Vaccination policy • Restrictiveness • Coercion • Least restrictive alternative

© The Author(s) 2019   59
A. Giubilini, *The Ethics of Vaccination*, Palgrave Studies
in Ethics and Public Policy,
https://doi.org/10.1007/978-3-030-02068-2_3

THE PRINCIPLE OF LEAST RESTRICTIVE ALTERNATIVE IN PUBLIC HEALTH

In the last chapter, we saw how there is a collective responsibility to realize herd immunity against vaccine-preventable infectious diseases, an individual responsibility to make one's fair contribution to the realization of herd immunity, and an institutional responsibility to implement vaccination policies that *at the very least* guarantee the realization of herd immunity.

Now, there are different types of vaccination policies that could be successful in realizing herd immunity, depending on factors such as particular socio-economic circumstances or cultural contexts. In order to decide which policy to implement among the potentially effective options, it is commonly acknowledged that policymakers ought to adopt principles of least infringement and of least restrictive alternative.

The principle of least infringement is a central pillar of public health ethics (Childress et al. 2002, p. 173). The principle states that public health authorities, when choosing between available policies for achieving a certain public health goal, should select the health policy that infringes the least upon certain individual rights. Such rights include the right not to be harmed, the right to receive beneficial medical treatments, the right to free movement and association, and the right to bodily integrity and to personal autonomy. In particular, with regard to bodily integrity and limitation of autonomy, which are the two prima facie rights that coercive vaccination policies seem to threaten (either parental autonomy in the case of child vaccination or individual autonomy in the case of competent individual vaccination), the principle of least infringement gives rise to a *principle of least restrictive alternative* (PLRA) (Childress et al. 2002, p. 173). The PLRA can be stated as follows: "if two interventions can both efficaciously and effectively address a public health or health policy issue and are equal in all other morally relevant respects, the intervention least restrictive of personal liberties ought to be preferred" (Saghai 2014, p. 350). According to Lawrence Gostin, the PLRA requires implementation of the policy that entails "the least intrusion on personal rights and freedoms" whilst being capable of achieving the relevant public health goal (Gostin 2008, p. 142).

In line with the PLRA, the Nuffield Council on Bioethics has formulated an "intervention ladder" that ranks possible public health measures according to their degree of restrictiveness of individual autonomy. At the bottom of the ladder, we find interventions such as providing people with information about healthy practices, while at the top, we find maximally restrictive interventions such as restriction of choices (e.g., removing unhealthy ingredients from food) and outright compulsion (Nuffield Council on Bioethics 2007, pp. xviii–xix). In this chapter I will focus on the problem of identifying the least restrictive yet effective alternative for vaccination policies—which, for the moment, I will assume should aim at herd immunity, in accordance with the argument of the previous chapter. The restrictiveness of any type of intervention depends, among other things, on variables such as the psychology of the individuals targeted by a certain public health measure or their socio-economic circumstances. For example, giving financial incentives to parents for vaccinating their children might exert a different influence on the decision-making of different individuals, depending on the extent to which they are in need of money. For some people, an incentive may be impossible to reasonably refuse while others might remain indifferent to the incentive, thus maintaining their autonomy of choice. To give another example, the level of autonomy restriction of mandatory vaccination policies that make vaccination a requirement for enrolling children in public day care or school might depend on whether parents can afford and are willing to pay for home schooling.

The different influence of different possible policies on the decision-making of different individuals also suggests that the degree of effectiveness of any policy in achieving a certain public health goal is context-dependent. A systematic review of studies concerning different possible strategies to address vaccine hesitancy concluded that, in order to be effective, strategies should be tailored to the characteristics of the targeted populations, such as the specific reasons for hesitancy and the socio-economic context (Jarrett et al. 2015). For instance, we can hypothesize that information campaigns would be more effective where parents are concerned about the risks of vaccine side effects on their children, which is one of the most common reasons for vaccine refusal in the US (Salmon et al. 2005), even if, as we will see below, some evidence suggests that information by itself is less effective that one might initially think (Nyhan et al. 2014). In any case, information campaigns are (even) less likely to be effective in the case of vaccine refusals motivated by mistrust in health institutions or health professionals, which is more common in Europe (Yaqub et al. 2014), or in the case of refusal motivated by religious beliefs. Similarly, some forms of nudging, such as vaccinating children at school by default and allowing parents to opt out if they so wish, would be more effective where parents do not vaccinate their children merely because of the inconvenience that vaccination normally entails (such as having to pay a visit to the doctor). But, once again, nudging is likely to be less effective in the case of parents with deeply held religious or philosophical beliefs against vaccination, for example, a commitment to "natural" lifestyles (whatever this means). Thus, we would need to adopt different solutions in different contexts in order to find the policy that, consistently with the PLRA, is the least restrictive alternative that is also effective at realizing herd immunity.

Appealing to the PLRA in the case of vaccination policies presupposes the existence of an intervention ladder like the one provided by the Nuffield Council on Bioethics, with specific child vaccination policies ranked from the least to the most restrictive. However, there is a lack of discussion in public health ethics explicitly aimed at providing such a ranking. By contrast, the PLRA has been widely discussed in the context of mental health law and ethics (e.g., Johnston and Sherman 1993; Miller 1982), where the issues addressed have included the permissibility of confining mentally ill individuals in order to protect them and the community at large, as well as whether and to what extent it is permissible to enforce behaviour-changing methods for such individuals. While the aims and scope are different, some lessons might be learnt from the discussion in that field. For example, as Johnston and Sherman (1993) have argued, it is widely acknowledged within mental health law that other, less intrusive procedures must first have been shown to be ineffective before a more intrusive procedure can be implemented (Johnston and Sherman 1993, p. 106). It seems reasonable to suggest that, if we endorse the PLRA, vaccination policies should follow the same logic. Therefore, an intervention ladder based on restrictiveness of different vaccination policies is needed in order to allow policymakers to try different policies starting from the least restrictive ones. This chapter aims to provide just such an intervention ladder for vaccination policies.

Now, one might wonder whether it is even possible to rank vaccination policies according to their restrictiveness. After all, as I have said above, the degree of restrictiveness of different possible policies is context-dependent. Also, what criteria should be used to determine the position on the ladder of any policy? Ideally, given the ineliminable degree of uncertainty, the most plausible answer is that policies should be preferred, other things being equal, if they are (1) likely to be restrictive for the smallest population possible *and* (2) likely to exert the lowest degree of restrictiveness possible for that population. But this answer, by itself, is far from being satisfactory, given that the two criteria might be in conflict with one another. I will address this difficulty in the next section. Having laid the conceptual foundations for my analysis, I will then proceed by introducing the concept of coercion, which can be applied to some types of policies and can be used to assess their level of restrictiveness. After that, I will provide an intervention ladder of possible vaccination policies, each of which is discussed in a separate section of this chapter. I will suggest that public health authorities should take this ladder as a guide for implementing effective vaccination policies in order to comply with the PLRA.

RESTRICTIVENESS AS AUTONOMY VIOLATION AND THE CRITERIA FOR MEASURING IT

It seems reasonable to measure restrictiveness of vaccination policies in terms of level of infringement of individual autonomy that a certain policy entails. The reason is that people who are opposed to vaccines or who for any reason do not want to vaccinate themselves or their children often appeal to their autonomy to justify their choice, and they typically oppose vaccination policies that, in different ways and degrees, force them to vaccinate by claiming that such policies infringe upon their autonomy—either bodily autonomy or parental autonomy. While "autonomy" is a philosophically problematic concept, here I will understand autonomy simply as "the control an individual has over his or her own evaluations and choices" (Hausman and Welch 2010, p. 128). This conception of autonomy seems closer to what those who are opposed to vaccines or are sceptical about their benefits claim is violated when they are forced to vaccinate themselves or their children.

We have seen above that there are two criteria for measuring the restrictiveness of possible child vaccination policies. These are the likelihood (1) that a certain policy will be restrictive for the smallest population possible *and* (2) that the policy would exert the lowest degree of restrictiveness possible, compatibly with a sufficient degree of effectiveness. But the two criteria might be in tension with one another. Policies that are likely to be restrictive, that is, autonomy-infringing, for a greater number of people might infringe upon the autonomy of the affected individuals less than policies that are restrictive for less people. Consider, for example, nudging in the form of making school-administered vaccination the default option and giving parents the possibility to opt out. This type of child vaccination nudging could limit the autonomy of a greater number of people than would incentives for vaccinating one's own children. The reason why nudging limits autonomy (understood by its aforementioned definition) is that almost everybody is subject to the same biases that cause one to bypass autonomous and rational decision-making and hence makes nudging effective, as we will see in a later section. By contrast, financial incentives would only restrict the autonomy of the very poorest in society, for whom such incentives would amount to an offer that is simply "too good to refuse". However, the restriction of autonomy exerted by incentives is arguably greater than the restriction of autonomy entailed by nudging, in terms of magnitude of influence on individuals' decision-making. On the one hand, there are offers that the poor might simply find too good to refuse no matter how deeply held their anti-vaccination beliefs are: the influence of incentives on the poor's decision-making in such cases is significant. On the other hand, as we shall see, people with deeply held beliefs against vaccination probably have the cognitive resources to overcome the cognitive biases exploited by nudging. Therefore, they are likely to preserve their capacity for autonomous choice in spite of the nudging.

But how, then, can we rank policies on the basis of their degree of restrictiveness, if the two more plausible criteria for measuring restrictiveness can yield different results? What criterion should be given priority in formulating a ranking that could provide ethical guidance for public policy: the number of people who are likely to experience infringements of autonomy or the degree of autonomy infringement experienced, even if by fewer people?

I propose that we should adopt a combination of the two criteria. More precisely, we should prefer the policy that infringes the least upon the autonomy of any individual, unless the number of people who experience a lesser degree of autonomy violation is sufficiently large to morally outweigh the consideration of the higher degree of autonomy violation that would otherwise be experienced by those who are worse off. In other words, I suggest the adoption of the *maximin* criterion for the distribution of the burdens of a certain policy, constrained by a utilitarian calculus based on the consideration of the number of people who are burdened by a certain policy. The combination of these two criteria seems in line with some ethical intuitions that most of us would share. Let us see more in details.

Our purpose is to formulate a ranking that can provide *ethical* guidance. This means that "restrictiveness" is not only a descriptive but also a normative concept: policies that are less restrictive *ought to* take priority over policies that are more restrictive. Therefore, when two descriptive criteria for determining the degree of restrictiveness conflict with one another, normative considerations about what ought to be done are relevant in determining what criterion ought to prevail in determining the degree of restrictiveness. The criterion that tells us which policies are less restrictive than others would also tell us which policies are ethically preferable to others. What are these normative considerations?

The two fundamental ethical requirements on which most reasonable people would probably agree seem to be exactly the two criteria mentioned above, namely, that (1) individuals should be burdened to the lowest degree possible, compatibly with the effectiveness of any given policy, and that (2) the total number of individuals burdened by a certain policy should not be too large. The two criteria can be combined in the sense that there must be some point beyond which, intuitively, the number of individuals burdened is so large that it outweighs the magnitude of the burden experienced by the worse off in terms of autonomy violation. Thus, policies that burden individuals less ought to be preferred to—that is, are to be considered less restrictive than—policies that burden individuals more, unless the number of individuals who are burdened less than others is sufficiently large, in which case the policy that burdens individuals more is to be considered less restrictive and therefore is to be ethically preferred.

For instance, to consider an extreme case, suppose we are choosing between two different policies that will affect one million people. Further suppose that we can measure restrictiveness on a scale 0–100, where 0 indicates no restrictiveness at all and 100 the highest degree of restrictiveness. Policy A restricts 1 person's choices to a degree of 50 and restricts the choices of 999,999 people to a degree of 0; meanwhile, policy B restricts the choices of all 1 million people to a degree of 49. It seems implausible that we should prefer B, even if the burden on any individual in policy B is lesser than the burden on one individual in policy A. The least restrictive policy is in this case policy A. Thus, my suggestion is that, for our purposes, the least restrictive policy, and therefore the policy that ought to be preferred, is the one that restricts the least the autonomy of those who are worse off in terms of autonomy restriction—according to what Rawlsians would call the *maximin rule* (Rawls (1971) 1999,

p. 133)—up to the point at which the number of those who experience some level, even a lower level, of autonomy restriction becomes sufficiently high. It follows that we should care somewhat about fairness in the distribution of restrictiveness across people and somewhat about total restrictiveness (i.e., degree of restrictiveness × number of people restricted).

However, in ranking possible vaccination policies on the basis of their restrictiveness, we need to have one clear criterion in mind. In what follows, I will adopt the maximin criterion as the primary criterion: I will rank the possible vaccination policies, from least to most restrictive, on the basis of how restrictive they are likely to be for those who are more significantly restricted by the policy in question. (The more precise meaning of "being restricted" will be discussed in the next section.) The choice is motivated not by some specific normative theory, but simply by an intuition I have, and which I think most people would have, when thinking about a fair distribution of certain burdens: it seems to me that we should prioritize placing the smallest possible burden on the worst off and that we should then constrain this criterion only by ensuring that not too many people are significantly burdened in order to protect the worst off. The intuition might be mistaken, but it seems to be supported by approaches to distributive justice that are normally considered reasonable, such as the one based on Rawls' famous "veil of ignorance", adjusted through utilitarian considerations.

The utilitarian constraint means that the maximin criterion I have adopted only provides a *provisional* ranking. It is understood that, in accordance with the combination of the two criteria, the ranking would have to be modified in case a certain policy that exerts a lower degree of restrictiveness on the worst off is likely to negatively affect (in terms of restrictiveness) a significantly larger number of individuals than a different policy. Thus, for example, incentives can affect the capacity for autonomous decision-making of some individuals more heavily than nudging because, as we mentioned above and as we shall see in more details below, it can be easier to counteract the psychological mechanisms exploited by nudging than it is to resist the temptation to accept an incentive. For this reason, nudging comes before incentives in my intervention ladder. However, in cases where only a very small part of the affected population is in such a poor socio-economic situation that they cannot refuse incentives, or if the number of people who are affected by nudging is sufficiently large, we would need to change the order and rank nudging after incentives. When and where this is the case depends on factors that are context specific.

Marcel Verweij and Angus Dawson have proposed that participation in collective vaccination programs (including child vaccination) should be *voluntary*, unless *compulsion* is necessary to prevent serious harm (Verweij and Dawson 2004). Voluntary and compulsory vaccinations constitute the two extremes of the ladder, involving the minimum and maximum degrees of restrictiveness, respectively. However, one problem with drawing this type of dichotomy is that, between compulsory vaccination and voluntary vaccination, there is a spectrum of different possible interventions involving different degrees of restrictiveness. For instance, the Italian government recently decided to follow the example of the US in making certain vaccinations mandatory, as complying with vaccination schedules has become a requirement for enrolling children in state-sponsored nurseries or preschools. As we will see more clearly after the discussion in the next section, this is an example of a position involving *some* coercion, which therefore is more coercive (and more restrictive) than completely voluntary vaccination, whilst being less coercive than outright compulsion: parents remain free not to vaccinate their children, although, in practice, such choice has a cost that constrains their autonomy. Predictably, only some parents would be able to afford private day care, and presumably even fewer would be willing to pay for it even if they could afford it. But in what sense we can say that this policy is somewhat "coercive"? I turn to this question in the next section.

RESTRICTIVENESS AND COERCION

Before presenting the intervention ladder, it is useful to say something more about coercion, given that some vaccination policies are—or at least are often referred to as—coercive. Since people often claim that it is wrong for a state to coerce them into vaccinating themselves or their children, let us examine what it means for a policy to be coercive and why and to what extent coercion in vaccination policies might be thought to be ethically wrong. The notion of coercion has a long philosophical tradition, and some insights from this philosophical debate can shed light on the conceptual and normative implications of restrictiveness.

Many different definitions of coercion have been proposed in the philosophical literature, and the notion has several different meanings in everyday language (Wertheimer 1989, pp. 185–188). Alas, a comprehensive overview of these definitions and meanings is beyond the scope of this chapter. For the purpose of the present discussion, we can follow those authors who define coercion in psychological terms, that is in terms of *influence* of a certain proposal (or policy) on a person's will (e.g., Frankfurt 1973; Feinberg 1989). More specifically, coercion can be conceived as a condition in which someone is forced to do X, for example, vaccinating one's children, in the sense that she is left with "no reasonable choice" or "no acceptable alternative" (Wertheimer 1989, pp. 30, 36–37) but to do X when she would otherwise not choose to do X. In other words, in cases of coercion a person's *autonomy* is infringed upon in a certain specific way, i.e. by making certain choices unreasonable or unacceptable, and by subjecting her will to the will of another (Frankfurt 1973, p. 80), where this "other" might be a state. Coercive interventions thwart autonomy—as I have defined it above—to the extent that they render unreasonable those choices that individuals would otherwise make on the basis of their own evaluation. Importantly, on the account of coercion I endorse, someone could be coerced into doing X not only by a proposal that attaches penalties to not doing X—that is, a threat, for example, excluding unvaccinated children from school—but also by a proposal that attaches significant enough benefits to doing X—that is, an offer (Held 1972; Feinberg 1989; Frankfurt 1973), for example, giving very large financial incentives for vaccinating one's children.

Thus, the definition of coercion I have provided differs from "baseline accounts" of coercion. According to these, what is relevant for the definition of "coercion" is the distinction between threats and offers, as defined by prospected changes with regard to a certain baseline. The idea behind baseline accounts is that coercion necessarily involves a threat, and offers can never be coercive (e.g., Nozick 1969; O'Neill 1991; Wertheimer 1989; Beauchamp and Childress 2001, p. 95). According to Nozick, one difference between threats and offers is that only the latter preserve freedom; that is, "when someone does something because of offers it is his own choice, whereas when he does something because of threats it is not his own choice but someone else's" (Nozick 1969, p. 459). This view, however, overlooks the influence on individual decision-making that very appealing offers can have. The account I endorse takes instead such influence into consideration. In some cases, for example, when the recipient desperately needs money, offers can leave the recipient with no reasonable choice but to accept what is offered, for example, a financial incentive, and to comply with the conditions of the offer, for example, vaccinate their children. In this sense, we cannot exclude that a certain offer might constitute a form of seduction (Held 1972) to which it is difficult or impossible not to succumb, although it is true that generally speaking the degree of coercion would often by much lower in the case of incentives than of penalties.

Also included in the notion of "coercion", as I will understand it, is that insofar as an individual is prevented from exercising her free will and judgement, coercion is pro tanto morally wrong. Accordingly, a moral justification that outweighs the prima facie wrongness of coercion is necessary in order to permissibly implement coercive public policies. One example of countervailing moral justification might be the realization of a public good like herd immunity. Admittedly, its positive value can trump the negative value of infringing upon certain autonomy rights of individuals. To be clear, my position is different from moralized accounts of coercion, according to which a proposal must by definition, in order to be coercive, threaten the recipient with the prospect of a wrongful action (Wertheimer 1989, p. 30)—as in "your money or your life" (where, for fear of stating the obvious, killing is the prospected wrongful action). On these accounts, coercion is prima facie morally wrong independently of the fact that it infringes upon autonomy (although the autonomy infringement in case the recipient accepts the proposal would add to the wrongness of the proposal). Instead, the reason why I consider coercion pro tanto wrong is precisely the fact that it infringes upon autonomy together with the consideration that we have a pro tanto moral reason for respecting individuals' autonomy.

Although coercion certainly makes a vaccination policy restrictive, it is important to point out that a policy can be restrictive without being coercive. The ranking I am going to propose takes into account factors other than coercion, because there are non-coercive ways of restricting individual autonomy: a policy can restrict individual autonomy without leaving individuals with "no reasonable choice" or "no acceptable alternative". In other words, the notion of restrictiveness is broader than that of coercion. For example, someone can be restricted in a non-coercive way if her capacity for autonomous decision-making is circumvented through nudging or by exploiting some cognitive bias. Thus, appeals to the notion of coercion will help us in drafting our ranking only with regard to the relative positions of those policies that are both coercive and restrictive.

Restrictiveness also depends on another factor, unrelated to the degree of coercion or of autonomy infringement, namely, *what* a person is forced to do. For example, it seems intuitively plausible to say that being coerced to have one's children vaccinated is less restrictive than being coerced to, say, donate one's kidney. However, while this consideration is important in a comprehensive conceptual analysis of restrictiveness, it is not relevant for the purpose of compiling a ranking of vaccination policies on the basis of restrictiveness since, with any policy, the autonomy right being restricted remains constant, namely, the right to make autonomous decisions over one's body or one's child health.

In the next sections, I am going to present my proposed intervention ladder. I will introduce and discuss the different possible child vaccination policies from the least to the more restrictive. I will start with the least restrictive non-coercive type of policy, namely, persuasion.

PERSUASION

Let us start with what we might call level zero of restrictiveness or coerciveness: mere persuasion. Some form of persuasion in public health communication, such as education campaigns to promote vaccination uptake, might be deployed to encourage people to vaccinate their children. Persuasion is a type of communication that aims at influencing individuals' behaviour (Rossi and Yudell 2012, p. 192). In the context of public health, persuasion has been defined as a "form of interpersonal influence, in which one person tries to change the attitudes or behaviour of another by means of argument, reasoning, or, in certain cases, structured listening". (Warwick and Kelman 1973, quoted in Faden and Faden 1978, p. 183), or in which "a person comes to believe in something through the merit of reasons another person advances" (Beauchamp and Childress 2001, p. 94).

Despite its being aimed at influencing individual behaviour, a distinguishing feature of persuasion so understood is the fact that it is both non-coercive and non-manipulative. By contrast, manipulation infringes, to a certain extent, upon individuals' autonomy by bypassing their capacity for autonomous decisions (Rossi and Yudell 2012, pp. 193–194). For example, manipulation might use subliminal messages or enlist community opinion leaders as allies in pro-vaccination campaigns (Colgrove 2016, p. 1316) or, as we shall see in the next section, deploy some form of nudging. Mere persuasion, on the other hand, preserves individuals' autonomy by relying merely on provision of factual information and of reasons for engaging in a certain behaviour. This means that individuals generally maintain the capacity to overcome the influence to which they are subjected. I might be exposed to messages concerning the safety and benefits of vaccines, which provide me with pro tanto reasons to vaccinate my children; however, if my anti-vaccination beliefs are deeply held or my anti-vaccination sentiments are strong enough, I would probably maintain my capacity to make an autonomous decision not to vaccinate my children, in spite of such messages. Accordingly, I place persuasion at the bottom of my intervention ladder.

Following Stanley Benn, Faden and Faden (1978, p. 186) use the concept of "persuasion" to refer both to persuasion as we have defined it above and to manipulation. However, they maintain the conceptual distinction between the two by distinguishing between "rational" and "non-rational" persuasion. While the former is based on the strength of substantial arguments, the latter aims at influencing individuals' behaviour by bypassing their capacity for rational thinking, for example, through the manner or style in which the arguments are presented. Contrary to what Faden and Faden (1978, p. 188) argue, non-rational persuasion is not coercive, at least according to the definition of coercion I have provided above, because it is incorrect to say that it leaves individuals "with no reasonable choice" but to pursue a certain course of action. However, even if not coercive, non-rational persuasion is manipulative and fails to protect autonomy of choice. This does not necessarily mean that non-rational persuasion, or manipulation in general, is morally unjustifiable: individual autonomy is only one value among many others in public health. The public interest in having enough individuals vaccinated might justify the circumvention of individual autonomy in order to convince them to opt for vaccination. What matters for the purposes of the present discussion is that non-rational persuasion and manipulation circumvent individuals' rational deliberative process and are therefore more autonomy restrictive than rational persuasion. Thus, if we want to refer to persuasion as a form of public health intervention that lies at the bottom of our intervention ladder, that is, that exerts the lowest degree of restrictiveness possible, we need to refer only to rational persuasion. To introduce yet another equivalent concept, some have referred to what Faden and Faden call rational persuasion by using the term "health *education*", understood as "any combination of learning opportunities designed to facilitate *voluntary* adaptation of behavior which will improve or maintain health" (Green 1978). In the case of rational persuasion or education, the autonomy to choose whether or not to vaccinate one's children is preserved.

Whether rational persuasion or education would be effective in keeping child vaccination rates high, or in increasing child vaccination rates in any given context, is an open question. In an experiment, a group of hesitant parents were provided with different messages—including both images and verbal information—about the MMR vaccine safety and effectiveness, as well as the risks of the diseases targeted by the vaccine. None of the messages convinced parents to vaccinate their children, and in some cases even reduced vaccination intention and activated a post-hoc rationalization. As Nyhan and colleagues explained: "respondents brought to mind other concerns about vaccines to defend their anti-vaccine attitudes, a response that is broadly consistent with the literature on motivated reasoning about politics and vaccines" (Nyhan et al. 2014, p. 6). Besides, even if certain interventions are successful in increasing confidence in vaccines, it is unknown whether increased confidence has any impact on vaccination uptake (Brewer et al. 2017).

The effectiveness of rational persuasion is likely to depend on the reasons why parents would be inclined not to vaccinate. As we have seen in Chap. 1, the phenomenon of vaccine hesitancy is complex, and in any given cultural or socio-economic context, there might be different predominant reasons why people decide not to vaccinate. These include perception of risk, lack of trust in health professionals, or religious or personal moral reasons (Dubé et al. 2013). As put by the WHO's Report of the SAGE Working Group on Vaccine Hesitancy, "[v]accine hesitancy is complex and context specific, varying across time, place and vaccines. It is influenced by factors such as complacency, convenience and confidence" (WHO 2014, p. 8). Therefore, persuasion might work in certain contexts but not in others. A recent study has shown that in the US 74% of parents who refused to vaccinate their children believed that vaccines are unnecessary, while 64% were concerned about possible links between vaccination and autism and/or about the presence of thimerosal in vaccine shots (Hough-Telford et al. 2016)—both of which represent misplaced concerns. An older study showed that 69% of parents refusing vaccination for their children were concerned that vaccines might cause harm in a more general sense (Salmon et al. 2005). These people seem to be the proper target of persuasion or health education campaigns. However, mere persuasion would probably not be effective in the case of parents with a religious or a philosophical opposition to vaccines. In such cases, policies with a higher degree of influence on individual decision-making might be required in order to realize herd immunity.

NUDGING

Moving on along our intervention ladder, we find a policy that is also non-coercive and minimally restrictive, although more restrictive than mere persuasion, namely, influencing people's choices through nudges. A nudge is a way of setting up the range of choices that "alters people's behavior in a predictable way without forbidding any option or significantly changing their economic incentives" (Thaler and Sunstein 2008/2009, p. 6).

Nudges exploit certain decision biases and automatic cognitive processes, harnessing them in order to encourage certain behaviours (Li and Chapman 2013, p. 188). In this way, nudges bypass some of people's deliberative capacities and therefore diminish people's capacity for autonomous decision-making. In other words, nudging is a manipulative strategy (Navin 2017, p. 47; Ploug and Holm 2015; Blumenthal-Barby and Burroughs 2012, p. 5). However, it is not a coercive strategy, since it does not leave individuals with "no reasonable choice" or "no acceptable alternative". In their seminal work on nudging, Richard Thaler and Cass Sunstein use the expression "libertarian paternalism" to describe the ethical framework that justifies the use of nudges. The "libertarian" aspect lies in the idea that people remain free to do what they like, in the sense that all the options remain open to them. The paternalistic aspect "lies in the claim that it is legitimate for choice architects to try to influence people's behavior in order to make their lives longer, healthier, and better" (Thaler and Sunstein 2008/2009, p. 5). Or, we might add, in the case of child vaccination, in order to protect the health of themselves, their children, and of the whole community.

One of the clearest cases of manipulation through nudging is the exploitation of status quo bias, that is, people's a priori preference for the status quo over possible alternatives (Thaler and Sunstein 2008/2009, p. 37). Status quo bias gives rise to a "default effect", that is, "the tendency for decision makers to stick with the default, or the option that takes effect if one does not make an explicit choice" (Li and Chapman 2013, p. 190). An example of the default effect is found in opt-out policies regarding organ donation, where people are presumed to consent to donating their organs after death unless they declare otherwise. Some evidence suggests that where opt-out policies are in place, organ donation rates are higher, thus showing the influence of the default effect on individuals' decision-making (Thaler and Sunstein 2008/2009, pp. 187–188). In the case of vaccination, nudges of this type might prove particularly effective in consideration of so-called literal inconsistency which is often found in vaccination decisions: parents with favourable vaccination intentions often do not act upon their intentions (Brewer et al. 2011, 2017). In such cases, nudging vaccination might simply be a way of removing those obstacles—whether psychological, material, or both—that prevent people from implementing their vaccination intentions.

For example, nudges could be implemented so as to exploit some of the very same decision-making biases that explain some people's refusal of vaccination and turn them into psychological mechanisms that orient individuals' choices towards vaccination. Opel et al. (2013) demonstrated the decisive role that the "default effect" plays in vaccination discussion between healthcare providers and hesitant parents in parents' vaccination decisions. In their study, they distinguished between presumptive formats of discussion, that is, formats "that linguistically presupposed that parents would vaccinate, such as declaration that shots would be given (e.g., 'Well, we have to do some shots')" (Opel et al. 2013, p. 3), and participatory formats, that is, formats "that linguistically provided parents with relatively more decision making latitude, such as polar interrogatives (e.g., 'Are we going to do shots today?') and open interrogatives (e.g., 'What do you want to do about shots?'), or ones that presupposed that parents would not vaccinate (e.g., 'You're still declining shots?')" (Opel et al. 2013, p. 3). The authors found that "a larger proportion resisted vaccine recommendations when providers used a participatory rather than presumptive initiation format" (83% vs 26%; $P < 0.001$) (Opel et al. 2013, p. 4). The authors concluded that "[h]ow providers initiate their vaccine recommendations at health supervision visits appears to be an important determinant of parent resistance to that recommendation" (Opel et al. 2013, p. 6).

But we might think of other ways to exploit the default effect in vaccination decisions. For example, children's vaccination in schools could become the default option. At the moment, in most countries, even when vaccination is a requirement for enrolling children in day care or schools, parents would normally have to actively authorize the vaccination and to pay a visit to the doctor. But by changing the default option, all the children enrolled in day care or school would be vaccinated, for example, by a doctor visiting the institution or by school nurses. Parents would not be asked for explicit consent, but they would be informed and given the option to opt out for their children if they so wish, in line with the idea that nudges should not forbid any option. By doing nothing, parents would be implicitly authorizing the vaccination of their children.

Some have argued that making the exemption procedure itself particularly burdensome from a bureaucratic point of view—for example, requiring notarization of forms, hand delivery, physician confirmation of information disclosure, and so on—would also represent a form of nudging (Lynch 2016, p. 110). This might be true, but seemingly only up to a point. Part of the concept of nudging is not only that decision makers retain their freedom of choice (Li and Chapman 2013, p. 188), but also that no option, including opting out, should be particularly costly for the chooser (Thaler and Sunstein 2008/2009, p. 5; Blumenthal-Barby and Burroughs 2012, p. 3).

One might object to the use of nudging by appealing to the Kantian categorical imperative, in its formulation that prescribes one to always treat other people also as ends in themselves and never merely as means. Since nudging would circumvent certain deliberative capacities and thus diminish individual autonomy, it would imply that people are treated not as ends in themselves, which would require respecting their autonomy, but as mere means to benefit other people. Now, it is true that, on a Kantian understanding of "means", nudging would imply treating individuals as mere means. However, there are two considerations that mitigate the wrongness of violating the categorical imperative.

First, nudging would often benefit not only society at large but also individuals who are nudged, given that vaccination offers a very high degree of protection against infectious diseases. Nudging in the interest of the those being nudged would make any limitation of autonomy more morally acceptable than nudging that is solely in the interest of the one doing the nudging (Halpern et al. 2007) or of third parties (such as society at large). If individuals are used as mere means, this will often be to their own benefit as well; and it is at least doubtful that autonomy violation represents such a serious wrong that it cannot be justified even by the large benefit it would entail to the individual in question.

Second, outside of a Kantian framework, but within a very reasonable perspective, whether autonomy is such an important value seems to depend on the extent to which making an autonomous choice in a certain context matters to an individual. If vaccination were the default option, since the possibility to opt out would remain open to them, parents would still be able to make the autonomous choice not to vaccinate their children in cases in which they have a *strong enough* desire to avoid vaccination, that is, in cases in which making an autonomous decision about vaccination matters to them. By "strong enough desire" I mean a desire that is sufficiently strong to overcome automatic cognitive processes, such as the a priori preference for the default option. The autonomy of parents with strong enough beliefs or other attitudes against vaccination would be preserved even if child vaccination were the default option. As explained by Yashar Saghai, there is sufficient psychological evidence to believe that "at least when individuals have strong enough preferences, goals, or beliefs, they are likely to become aware of an anomaly" (Saghai 2013, p. 489), that is, of a discrepancy between their conscious desires and what they are nudged to do. Such awareness would enable them to inhibit the automatic cognitive process that the nudging would otherwise exploit. Nudging would only affect the decision-making of parents with weak and trivial beliefs against vaccination. This group includes, for instance, parents who would otherwise not vaccinate their children because they do not have time, do not want to go through the inconvenience of paying a visit to the doctor, or simply think that their child is healthy enough and there is no need for vaccination. In fact, at least some non-medical exemptions to child vaccination are obtained for reasons of mere convenience. This is suggested by the fact that school-based immunization clinics have proven to be effective in increasing the number of fully immunized students (Wang et al. 2014, p. e80). But the violation of autonomy in such cases of weak and trivial preferences for non-vaccination does not seem morally significant. To these parents, making an autonomous choice regarding their children's vaccination is not seen as especially valuable, at least not enough to overcome their automatic processes. Since parents do not oppose child vaccination, nudging them by making vaccination the default option would, although autonomy-infringing, not be autonomy-infringing in a morally problematic way.

INCENTIVES

So far, I have discussed two strategies that lie at the non-coercive end of the spectrum of possible interventions to promote child vaccination. I have argued that persuasion does not involve any interference with autonomy and that nudging can involve interference with autonomy, but that when the latter does, the interference is not morally problematic. I now turn to examining a third possible strategy that, as I shall argue, is more restrictive than the two examined so far, in that it potentially implies some coercion for at least some individuals. This type of intervention is the provision of financial incentives, or *conditional cash transfers* (CCTs), for vaccinating oneself or one's children.

What has been called the "archetypical aim" of CCTs is to make certain options less costly and hence more accessible and salient to individuals (Grill 2017, p. 159). In this section, I am going to discuss what Faden and Beauchamp (1986, pp. 357–58) and Krubiner and Merritt (2017) call "unwelcome" offers. These are offers of incentives for engaging in actions that are in tension with individuals' desires and will. For example, an offer of incentives for vaccinating one's children represents an unwelcome offer for parents who are opposed to vaccination.

Now, provision of CCTs for healthy behaviours, including behaviours that promote public health, raise ethical issues on many levels (Lunze and Paasche-Orlow 2013; Marteau et al. 2009), including the design of CCTs schemes, their implementation, and their possible unintended consequences. Carleigh Krubiner and Maria Merritt have argued that in designing CCT interventions, policymakers should attend five types of considerations. These are the likelihood of bringing about the desired benefits, the risks and burdens involved, the receptivity of the intended beneficiaries and of communities, the attainability of the program (e.g., what kinds of barriers to compliance exist for the beneficiary population), and the indirect impact and externalities (Krubiner and Merritt 2017). Moreover, once implemented, CCT programs raise distinctive ethical issues. These include, among others, the potential of incentives for bribery (paying people to act against their wishes), coercion, paternalism, unfairness (it might be argued that people should not be paid to do what they ought to do anyway), and poor use of scarce financial resources (Marteau et al. 2009). Finally, other concerns arise with regard to the possible unintended consequences of CCTs, such as the "crowding out" of intrinsic motivation (Krubiner and Merritt 2017, p. 170).

For the purpose of this chapter, what matters is the kind of influence that incentives have on individuals' capacity for *autonomous* decision-making. In this sense, incentives, at least when they are sufficiently large, are coercive in a way that persuasion and nudges are not, but they are less coercive than imposing penalties or than compelling people to adopt a certain behaviour. Let us consider these two comparisons in order.

First, sufficiently large incentives are coercive in a way in which persuasion and nudges are not, at least according to the definition of coercion I have adopted. Sufficiently large incentives can undermine the decision-making processes of vulnerable individuals, particularly of those on a low income (Voigt 2017; Blumenthal-Barby and Burroughs 2012, p. 2). If these individuals are opposed to vaccination, unwelcome offers of sufficiently large incentives would provide them with strong reasons and motivation to do what they would rather not do, that is, vaccinate their children. In contrast, nudging, although it circumvents autonomous decision-making, does not provide parents with any *reason* for vaccinating their children. What makes sufficiently large incentives morally problematic in a way that nudging is not is that, for certain parents, such reasons might be too strong to be disregarded. Thus, sufficiently large incentives can interfere with an individual's capacity for autonomous decision-making in a way in which persuasion and nudging cannot, that is, by leaving people with no reasonable choice or no acceptable alternative.

And indeed, as I have mentioned above, according to philosophical understandings of coercion that align with our definition (e.g., Held 1972; Frankfurt 1973), not only threats (i.e., proposals to make a person worse off if the person does not do X) but also offers (i.e., proposals to make a person better off if the person does X) can exert a coercive influence on an individual's will, at least when they are sufficiently large. According to Harry Frankfurt, just like a threat, an offer may "arouse in the person who receives it a desire—i.e. to acquire the benefit—which is similarly irresistible. This suggests that a person may be coerced by an offer as well as by a threat" (Frankfurt 1973, p. 79). And as put by Virginia Held, "as an inducement to accept an offer approaches a high level, it approaches coercion proportionately" (Held 1972, p. 57).

Of course, incentives become more coercive in proportion to their size relative to the economic circumstances of the recipient. For example, for the vast majority of people in developed countries, a small incentive of US$5 would be minimally coercive and less autonomy-restricting than nudging. In the developed world, even a US$50 incentive, or an incentive in the form of food or medicine coupons, might not have a significant impact on the decision-making of wealthy parents who are sceptical of vaccine efficacy or safety; however, the same incentive might be irresistible to sceptical parents on a low-income or in low-income countries. With regard to the degree of coercion involved by certain offers, Frankfurt (1973) has argued that when an individual A is dependent on B (another individual or a state) for a certain good (such as money), when A *needs* the good, and when B offers the good to A exploiting B's dependence and need, withholding a benefit, such as an incentive, is tantamount to imposing a penalty, that is, an offer is ethically and psychologically equivalent to a threat (see also Faden and Beauchamp 1986, p. 358). This view is quite extreme and ultimately incorrect, because, as I will argue in the next section, penalties are in an important sense more coercive than offers. However, milder versions of the same claim seem at least plausible: to certain individuals, coercion via significantly large offers is closer to the level of restrictiveness entailed by threats than it is to the level of restrictiveness entailed by persuasion or nudging (although it remains true that small offers might be easier to resist than nudges).

Thus, incentives are the first genuinely coercive form of intervention that we encounter as we move up on the intervention ladder. It is important to point out that coerciveness does not make incentives, all things considered, morally impermissible. It might still be justifiable to coerce individuals for the sake of a public good like herd immunity. However, the positioning of incentives on the intervention ladder after persuasion and nudging provides a pro tanto reason against incentives for vaccinating children—grounded in the PLRA—that is stronger than the pro tanto reason against persuasion and against nudging. Accordingly, *assuming* our aim is merely to realize herd immunity (and this is a proviso worth remembering, and that I will question in Chap. 4), incentives should be used only when persuasion and nudging turn out to be ineffective in realizing herd immunity and when there are sufficiently strong reasons, such as the public interest in realizing herd immunity, that outweigh that pro tanto reason. As far as coercion and restrictiveness are concerned, incentives are more ethically problematic than nudging or persuasion.

However, at the same time—and this is the second comparison I mentioned earlier—incentives are, other things being equal, ethically preferable to penalties for not vaccinating one's children, because they are less coercive according to my definition of coercion. In the next section, I am going to explain why three different types of penalties are more coercive than incentives.

Now, the relevant question is: are incentives effective? And if not, is it necessary to implement a more coercive type of policy, such as the imposition of penalties? Answering this question is complicated. Different systematic reviews have found conflicting and inconclusive evidence regarding the effectiveness of incentives in boosting vaccination rates, with regard to both child (Wigham et al. 2014) and adult (Lagarde et al. 2007) vaccination. It is difficult to draw general conclusions even regarding the effectiveness of incentives in promoting vaccination uptake in *types* of socio-economic contexts: conflicting evidence regarding incentives' effectiveness has been found both *within* high-income (Wigham et al. 2014) and *within* low- and middle-income (Lagarde et al. 2007; Ranganathan and Lagarde 2012) countries.

The size of the incentive obviously affects individual responses. A theoretical epidemiological game model based on a questionnaire about people's perceptions of influenza and of vaccines found that, in developed countries, in order to have a sufficient proportion of the population vaccinated against influenza, "providing incentives to encourage vaccination is inevitable" (Yamin and Gavious 2013, p. 2668); the same study found that "socially optimal incentives to the vaccinated individuals should be as high as US$57" (Yamin and Gavious 2013, pp. 2668–9). But also individual socio-economic background, culture, and religion might determine the extent to which incentives influence parents' decision-making regarding their children's vaccination. For example, parents who are opposed to vaccination for deeply held religious reasons would probably be insensitive to financial benefits. However, someone without any principled opposition to vaccines and from a low socio-economic background might see a financial incentive as a too tempting inducement. For this reason, she might decide to vaccinate their children even if she would otherwise not have done so.

Besides, incentives might determine a "crowding out" of intrinsic motivation for the incentivized option or might cause parents to believe that the incentivized option is uncommon or not in line with social norms, since—so individuals might think—people need to be paid in order to be convinced to choose that option (Gneezy et al. 2011; Grill 2017). It is however worth noting that one of the aforementioned systematic reviews about preschool vaccination uptake in high-income countries that claims to have found insufficient evidence to conclude whether financial incentives are effective (Wigham et al. 2014) included both a study about incentives understood as positive rewards (in the form of cash lottery tickets) and studies about incentives understood as avoidance of penalties (in the form of avoiding the withholding of certain state benefits): the conflicting evidence was found only with regard to the latter type. In fact, the only study included about positive rewards in the form of cash lottery tickets clearly showed that positive rewards, that is, genuine incentives, are effective, yielding a 21% increase in the number of vaccinations received by preschool children. Similarly, some evidence exists that suggests that incentives might be effective in middle- and low-income countries. For example, a study found that incentivizing child DTP vaccination (diphtheria, tetanus, and pertussis vaccines combined) with food coupons significantly increased vaccination uptake in a low-income area of Pakistan (Chandir et al. 2010).

Thus, in light of this evidence, we can conclude that financial incentives for vaccinating one's children *might* be effective, but whether they are is context-dependent and cannot be established a priori.

Where persuasion, nudging, and positive incentives are not effective, other more restrictive and more coercive policies would be required. One such type of policies is, as mentioned above, the imposition of penalties for non-vaccination, which I am going to discuss in the next section.

Disincentives

From a psychological perspective, the claim that threats of penalties are generally (though, as we will see, not always) more coercive than offers of incentives is supported by the existence of two related phenomena. The first is *loss aversion*, whereby "losing something makes you twice as miserable as gaining the same thing makes you happy" (Thaler and Sunstein 2008/2009, p. 36). In other words, when individuals have to give something up, such as money in the form of a fine for non-compliance with vaccination requirements, "they are hurt more than they are pleased if they acquire the very same thing" (Thaler and Sunstein 2008/2009, p. 36), for example, if they were to receive the same amount of money as an incentive for vaccinating their children. Loss aversion, in turn, explains the second, related psychological phenomenon known as *endowment effect*: we tend to value the goods that we already possess more than the goods that we do not possess yet, even if the good in question is the same or if the value of the good is the same (see e.g. Kahneman et al. 1991).

Thus, *generally* (and why I say "generally" rather than "always" will be clarified below), in virtue of loss aversion and of the endowment effect, threatening to impose a penalty for non-vaccination is more coercive than offering incentives for vaccinating one's children, other things being equal (e.g., the size of the incentive and of the offer is equally significant). In virtue of loss aversion and of the endowment effect, individual will is influenced to a greater extent by threats than by offers, in terms of creation of options that leave individuals with "no reasonable choice" or "no acceptable alternative".

Now, it is true that some authoritative studies have suggested that money, differently from other goods, does not create an endowment effect and loss aversion (Zamir 2015, p. 23; Novemsky and Kahneman 2005). However, other studies clearly show endowment effect and loss aversion generated by money and therefore by the threat of financial penalties. For example, a well-known experiment has shown that "framing teacher incentive programs in terms of losses rather than gains leads to improved student outcomes" (Fryer 2013). In this study, teachers were more strongly motivated to perform well if the prospect was losing money for poor performance rather than gaining money for good performance. As the authors of the study concluded, "there may be significant potential for exploiting loss aversion in the pursuit of (…) optimal public policy" (Fryer 2013).

And indeed, one type of public policy that could exploit the same psychological mechanisms is one based on penalties, or disincentives (for the purposes of the present discussion, I will take the two concepts as synonymous). This claim is further supported by the fact that among the two types of vaccination policies we are comparing—those based on incentives and those based on penalties—the difference between gains (incentives) and losses (financial penalties) is not merely a matter of framing, as was the case in the aforementioned teachers experiment; rather, it involves actual gains and actual losses. In the experiment above, teachers were differently influenced in their performance according to whether they perceived to be gaining or losing money, where the latter option exerted stronger influence. In the same way, it seems plausible to assume that parents would be influenced differently in their decision whether to vaccinate their children depending on whether they would receive an incentive for vaccination or would be required to pay a penalty for non-vaccination, where the latter option exerts stronger influence.

This difference between the perception of threats and perception of offers is reflected in the different philosophical and ethical treatment reserved to the two in philosophical discussions. From an ethical standpoint, and indeed within a common-sense perspective, threats are generally considered a bad thing, while offers are generally considered a good thing (Hetherington 1999, p. 211). Besides, as Harry Frankfurt noted, threats are generally thought to require justification, whereas offers are generally not (Frankfurt 1973, p. 83), although this claim is, admittedly, quite controversial (but we can leave the issue aside for the purpose of the present discussion).

One clarification is in order. I have said that threats are *generally*, rather than always, more coercive than offers. In fact, the actual relative influence of threats and offers on individuals will depends on the size of the prospected penalty or of the prospected benefit. The prospect of a very large incentive might exert a greater influence on many people's capacity for autonomous decision-making than the prospect of a very small penalty. Although I have placed penalties after incentives on the intervention ladder, the relative position of penalties and incentives might change when we compare significant incentives with less significant or insignificant penalties. Therefore, large enough incentives should come after small penalties on our intervention ladder, as the former are simply more coercive than the latter. Also, the amount of money that renders an offer coercive is probably not the same as the amount of money that renders a threat coercive. For example, because of loss aversion and endowment effect, an offer of a US$57 incentive for vaccinating their children might be less coercive for some parents than a threat of a US$57 penalty for not vaccinating their children. However, an offer of US$100 might be as coercive as a penalty of US$57, and an offer of US$150 might be more coercive than a penalty of US$57. Thus, we need to bear in mind that the amount of money that makes an offer significant is not necessarily the same as the amount of money that makes a threat significant.

Now, there are three main types of penalties, or disincentives, that can be imposed for non-vaccination. I am going to present them in the next three subsections in order of restrictiveness, starting with the least restrictive and least coercive one.

Withholding of Financial Benefits

Let us start with the withholding of financial benefits that the state would otherwise pay to parents. This type of policy has been introduced in Australia through the so-called "no jab, no pay" policy. Parents who do not comply with the recommended child vaccination schedule are no longer entitled to receive childcare benefits from the state. In that context, the policy has turned out to be effective.[1] However, once again, it is difficult to generalize from a single case. As mentioned earlier, a systematic review (Wigham et al. 2014) concluded that there is insufficient evidence for the claim that denying financial benefits is effective in increasing preschool vaccination uptake. As was the case with financial incentives, the effectiveness of the intervention might depend upon factors such as the

---

[1] See *The Guardian*, 'No jab, no pay': thousands immunize children to avoid family payment cuts, 31 July 2016, available at https://www.theguardian.com/australia-news/2016/jul/31/no-jab-no-pay-thousands-immunise-children-to-avoid-family-payment-cuts. Last accessed 26 May 2017.

particular socio-economic circumstances or cultural background of the targeted population.

The withholding of benefits is in many respects equivalent to the use of incentives. Most evidently, the financial benefit to which parents would be entitled if they vaccinated their children might be seen as an incentive for having one's children vaccinated. This justifies positioning the withholding of benefits close to incentives on our intervention ladder. However, withholding of benefits comes *after* and is therefore more restrictive and more coercive than the use of incentives because the withholding of financial benefits is analogous to the enforcement of a penalty with regard to its influence on individual decision-making. The reason is that at least some parents who decide not to have their children vaccinated would probably consider financial benefits to which they would otherwise be entitled as part of their normal baseline. Since at least some would perceive the financial benefit as part of the baseline, at least some of those who decide not to vaccinate their children might perceive the withholding of such financial benefits as a penalty, rather than perceiving the provision of the benefits as an offer. As far as restrictiveness and coerciveness are concerned, the withholding of financial benefits makes people who refuse vaccination not only worse off than they *would* be if they chose vaccination but also worse off than they probably *think they should* be. Thus, the withholding of financial benefits would, at least in some people, probably trigger the same loss aversion and the same endowment effect that is triggered by the imposition of financial penalties. Therefore, the withholding of financial benefits would be more coercive and restrictive than the provision of positive incentives, to which people do not think they are entitled. The PLRA implies that benefits should be withheld from non-vaccinating people only if persuasion, nudging, and the provision of incentives turn out to be ineffective in realizing herd immunity from a certain infectious disease.

Tax

It has been suggested that unvaccinated adults as well as parents who do not vaccinate their children for non-medical reasons should be subject to a financial penalty proportionate to the risk of infection the unvaccinated poses on other people (Clarke et al. 2017). According to the proponents of this view, the degree of risk is a function of the severity of the disease in question and of its morbidity.

Now, "financial penalty" can refer to two different things: either a tax or a legal sanction. The main difference between the two is that in the case of a tax, the behaviour which is taxed is legal, while a legal sanction means that the behaviour in question is illegal. When there is a legal sanction on non-vaccination, then vaccination is compulsory. Compulsory vaccination will be introduced in the last section of this paragraph and will be discussed in detail in the next chapter; here, I will briefly focus on the disincentive represented by a tax on non-vaccination. Taxing non-cooperative behaviour towards public goods would both discourage such behaviour and force people to internalize the costs of their failure to cooperate—similarly to how so-called Pigovian taxes force people to internalize the negative externalities, that is, the costs for society, of certain behaviours (e.g., drinking or smoking).

We can consider taxes for non-vaccination as a more coercive policy than withholding of incentives or financial benefits. As we have just seen, the reason why some people would consider the withholding of certain benefits as more restrictive and coercive than pure incentives is that such benefits might be taken to be part of the baseline that defines the status quo. With respect to such a baseline, withholding financial benefits represents a threat of a penalty and therefore makes the recipient worse off. On the basis of the same psychological account of coerciveness or restrictiveness, the level of coerciveness or restrictiveness people experience would be even greater if the sum of money claimed by the state is money that people actually already possess. With respect to the baseline, that is, the status quo, people would be even worse off after paying a tax than after not receiving a financial benefit. Therefore, the influence on autonomous decision-making exerted by taxes is greater than the influence exerted by the withholding of benefits. According to the PLRA, then, taxation for non-vaccination is a type of policy that should be implemented only if policies based on persuasion, nudging, incentives, or withholding of financial benefits turn out to be ineffective in realizing herd immunity. However, both withholding of financial benefits and taxes represent a form of penalty that is less restrictive and less coercive than the withholding of certain social services and goods, as we are going to see in the next section.

Mandatory Vaccination: Denying Enrolment in School and Day Care

A third type of penalty is non-financial in nature. Where mandatory vaccination is in place, children who do not comply with recommended vaccination schedules could be barred from enrolling in state schools or day care. Mandatory vaccination is enforced in some countries, for example, in the US and in Italy (in the latter case only for children of preschool age). In the remaining of this book, I will adopt the terminological distinction between *mandatory* and *compulsory* vaccination (Navin and Largent 2017). In the former case, but not in the latter, parents remain legally free not to vaccinate their children (e.g., by home schooling), although this choice comes at a cost that, as I will argue below, might be unreasonable for parents to choose to bear. This (conditional) freedom explains why mandatory vaccination comes before compulsory vaccination on our intervention ladder. At the same time, the fact that the choice not to vaccinate one's children has certain consequences that even wealthy parents might find extremely costly explains why mandatory vaccination comes after, that is, is more coercive than, withholding of financial benefits and taxation on the intervention ladder. In particular, parents would have to provide home schooling to their children or pay for private education.

Often, conscience clauses in vaccination legislations grant exemptions to this mandate on the basis of parents' religious or philosophical (depending on the legislation) opposition to vaccines. At the moment, all but three states in the US (California, West Virginia, Mississippi) have such conscience clauses in their legislations. Clearly, the existence of conscience clauses might render mandatory vaccination not only completely non-coercive but also ineffective. In this subsection, I will consider only mandatory vaccination that does not allow any conscientious objection, as is the case also in Italy. Where conscientious objection is granted, and especially if the exemption procedure is relatively easy and not burdensome, mandatory vaccination would boil down to, and therefore would not be any more restrictive than, a form of nudging.

Now, mandatory vaccination is similar in one important respect to the withholding of financial benefits that I have discussed in the previous section. Exactly like that policy, mandatory vaccination policies threaten to withhold a certain type of benefits from parents who do not vaccinate their children. However, mandatory vaccination is more coercive than the withholding of financial benefits because the type of good withheld does not have a merely monetary value. In the two cases of financial penalties examined above (withholding of benefits and taxes), people with sufficiently strong reasons against vaccination, a high socio-economic status, or both would have the option to make up financially for their vaccine refusal, perhaps by making great sacrifices if they are poor. But denying their children the opportunity to attend day care or school would impose a cost on both the parents (in case they cannot afford or do not want to provide private education) and, more importantly, their children for which it is more difficult, if not impossible, to make up. The cost is not merely financial, as there arguably are benefits associated with attending schools or day care in terms of children's well-being and healthy psychosocial development. Thus, unlike the case of financial penalties or of withholding of financial benefits, parents cannot fully compensate financially for the choice not to vaccinate their children. As a consequence, not only is the range of choices open to them more narrow than in the case of financial incentives or withholding financial benefits; also, and more importantly for the purpose of the present discussion, the choice not to vaccinate one's children would often be unreasonable, given that parents would probably be harming their children by denying them the same education that other children have. Therefore, these parents would often be left with "no reasonable choice" or "no acceptable alternative" but to vaccinate their children. This means that when they have to decide whether to vaccinate their children, they would experience a higher level of coercion than in the case of financial penalties in the form of taxation or withholding of financial benefits, that is, situations in which non-vaccination has a merely financial cost which parents might find reasonable to bear.

As before, the relevant question to ask is whether mandatory vaccination is effective. And once again, answering this question is complicated. In California, one year after the introduction of the legislation requiring all children without medical exemptions to have completed their recommended vaccination schedule in order to be enrolled in day care or school, 95.6% of kindergarteners ended up immunized for the school year 2016–17, up 2.8% from the previous year and the highest rate ever recorded in the state (California Department of Public Health 2017). However, a study on the effectiveness of state-level varicella vaccination mandates indicates that "the impact of the mandate is a *short-run* phenomenon. The importance of the mandate effect relative to the aggregate time trend (...) is cut by more than a half by the fourth year after the mandate and disappears completely approximately six to seven years after the mandate" (Abrevaya and Mulligan 2011, p. 971). Thus, where mandatory vaccination is not or no longer effective, a measure of last resort would be required, as I will briefly explain in the next section. However, as we shall see in the next chapter, there are other reasons in favour of compulsory vaccination policy that do not depend on the ineffectiveness of less restrictive policies to realize herd immunity.

COMPULSION

By making it illegal to refuse vaccination, compulsory vaccination would be the most restrictive and coercive type of vaccination policy. Compulsory vaccination would probably also be the most effective vaccination policy. In the next chapter, I will make the ethical case for compulsory vaccination, also by criticizing some of the authors who have put forward the idea that vaccination should be compulsory, but who have in the end mitigated their claim by appealing to the PLRA (Pierik 2016; Flanigan 2014).

That vaccination is compulsory means that there are *legal* penalties attached to non-vaccination. These legal penalties can range from a small fine to incarceration. In the next chapter, I will suggest that a significantly large fine is probably the most appropriate legal penalty for non-vaccination. In terms of costs incurred for non-vaccination, if we assume that the legal penalty should be financial in nature, compulsory vaccination is very similar to taxation for non-vaccination: in both cases we are imposing a financial penalty on those who fail to vaccinate. The difference is that, as I said above, in the case of taxation non-vaccination remains legal, while in the case of compulsory vaccination it is illegal. The difference is relevant because the fact that a certain option is illegal justifies imposing very burdensome legal penalties, while keeping the option legal means that the penalty should ideally remain within a reasonable range, since presumably a citizen should be put in the condition to autonomously choose between two legal options and therefore neither option should result too burdensome. Thus, compulsory vaccination is more restrictive than a tax on non-vaccination because compulsion allows an authority to enforce more substantial penalties that would exert a higher influence on individuals' capacity for autonomous decisions. Besides, the fact itself of breaking the law is likely to represent a strong psychological barrier to non-vaccination, as presumably many people would feel uncomfortable breaking the law and being subject to a legal penalty, which means that their autonomy would be way more affected than in case they were simply taxed. In the next chapter, I will provide an argument for compulsory vaccination.

For the purposes of this chapter, what matters is how compulsory vaccination fares in terms of consistency with the PLRA. Because compulsory vaccination is the most restrictive vaccination policy possible, it can only be consistent with the PLRA if all the other possible alternatives, from persuasion to withholding social goods or services, have proven to be ineffective in realizing herd immunity, and, of course, *only if we assume that the only aim of vaccination policies is the realization of herd immunity*. Thus, the PLRA implies that compulsory vaccination should be a measure of last resort, *if* what we want is simply the realization of herd immunity.

CONCLUSION

In this chapter, I have provided an intervention ladder that ranks possible vaccination policies on the basis of their degree of restrictiveness. *If* we accept that the principle of least restrictive alternative (PLRA) provides a reason to implement the least restrictive policy that is effective in realizing herd immunity, then governments have a reason to test the efficacy of possible policies in the following order, starting from the least to the most restrictive: persuasion, nudging, provision of incentives, withholding of financial benefits, imposition of financial penalties, withholding of social services and goods (e.g., enrolment in state school and day care; also known as mandatory vaccination), and, as a last resort, compulsory vaccination.

REFERENCES

Abrevaya, J., & Mulligan, K. (2011). Effectiveness of State-Level Vaccination Mandates: Evidence from the Varicella Vaccine. *Journal of Health Economics, 30*(5), 966–976.

Beauchamp, T. L. & Childress, J. F. (2001). *Principles of Biomedical Ethics.* Oxford: Oxford University Press.

Blumenthal-Barby, J. S., & Burroughs, H. (2012). Seeking Better Health Care Outcomes: The Ethics of Using the "Nudge". *The American Journal of Bioethics: AJOB, 12*(2), 1–10.

Brewer, N. T., Gottlieb, S. L., Reiter, P. L., McRee, A.-L., Liddon, N., Markowitz, L., & Smith, J. S. (2011). Longitudinal Predictors of Human Papillomavirus Vaccine Initiation Among Adolescent Girls in a High-Risk Geographic Area. *Sexually Transmitted Diseases, 38*(3), 197–204.

Brewer, N. T., Chapman, G. B., Rothman, A. J., Leask, J., & Kempe, A. (2017). Increasing Vaccination: Putting Psychological Science into Action. *Psychological Science in the Public Interest: A Journal of the American Psychological Society, 18*(3), 149–207.

California Department of Public Health. (2017). *2016–2017 Kindergarten Immunization Assessment – Executive Summary.* Retrieved April 2018, from http://eziz.org/assets/docs/shotsforschool/2016-17KindergartenSummaryReport.pdf

Chandir, S., Khan, A. J., Hussain, H., Usman, H. R., Khowaja, S., Halsey, N. A., & Omer, S. B. (2010). Effect of Food Coupon Incentives on Timely Completion of DTP Immunization Series in Children from a low-Income Area in Karachi, Pakistan: A Longitudinal Intervention Study. *Vaccine, 28*(19), 3473–3478.

Childress, J. F., Faden, R. R., Gaare, R. D., Gostin, L. O., Kahn, J., Bonnie, R. J., et al. (2002). Public Health Ethics: Mapping the Terrain. *The Journal of Law, Medicine & Ethics: A Journal of the American Society of Law, Medicine & Ethics, 30*(2), 170–178.

Clarke, S., Giubilini, A., & Walker, M. J. (2017). Conscientious Objection to Vaccination. *Bioethics, 31*(3), 155–161.

Colgrove, J. 2016. Vaccine Refusal Revisited—The Limits of Public Health Persuasion and Coercion. *New England Journal of Medicine 375*, 1316–1317. https://doi.org/10.1056/NEJMp1608967

Dubé, E., Laberge, C., Guay, M., Bramadat, P., Roy, R., & Bettinger, J. (2013). Vaccine Hesitancy: An Overview. *Human Vaccines & Immunotherapeutics, 9*(8), 1763–1773.

Faden, R. R., & Beauchamp, T. L. (1986). *A History and Theory of Informed Consent.* New York: Oxford University Press.

Faden, R. R., & Faden, A. I. (1978). The Ethics of Health Education as Public Health Policy. *Health Education Monographs, 6*(2), 180–197.

Feinberg, J. (1989). *The Moral Limits of the Criminal Law: Volume 3: Harm to Self.* New York: OUP.

Flanigan, J. (2014). A Defense of Compulsory Vaccination. *HEC Forum: An Interdisciplinary Journal on Hospitals' Ethical and Legal Issues, 26*(1), 5–25.

Frankfurt, H. (1973). Coercion and Moral Responsibility. In T. Honderich (Ed.), *Essays on Freedom of Action.* London: Routledge and Kegan Paul.

Fryer, R. G. (2013). Teacher Incentives and Student Achievement: Evidence from New York City Public Schools. *Journal of Labor Economics, 31*(2), 373–407.

Gneezy, U., Meier, S., & Rey-Biel, P. (2011). When and Why Incentives (Don't) Work to Modify Behavior. *The Journal of Economic Perspectives: A Journal of the American Economic Association, 25*(4), 191–210.

Gostin, L. 2008. *Public Health Law. Power, Duty, Restraint.* Revised and Expanded (2nd ed.). Berkeley/Los Angeles: California University Press.

Green, L. W. (1978). Determining the Impact and Effectiveness of Health Education as It Relates to Federal Policy. *Health Education Monographs, 6*(Supplement 1). Retrieved from https://eric.ed.gov/?id=ED183498

Grill, K. (2017). Incentives, Equity and the Able Chooser Problem. *Journal of Medical Ethics, 43*(3), 157–161.

Halpern, S. D., Ubel, P. A., & Asch, D. A. (2007). Harnessing the Power of Default Options to Improve Health Care. *The New England Journal of Medicine, 357*(13), 1340–1344.

Hausman, D. M., & Welch, B. (2010). Debate: To Nudge or Not to Nudge. *The Journal of Political Philosophy, 18*(1), 123–136.

Held, V. (1972). Coercion and Coercive Offers. In J. R. Pennock & J. W. Chapman (Eds.), *Nomos XIV: Coercion* (Vol. 14, pp. 49–62). Chicago: Aldine-Atherton.

Hetherington, A. (1999). The Real Distinction Between Threats and Offers. *Social Theory and Practice, 25*(2), 211–242.

Hough-Telford, C., Kimberlin, D. W., Aban, I., Hitchcock, W. P., Almquist, J., Kratz, R., & O'Connor, K. G. (2016). Vaccine Delays, Refusals, and Patient Dismissals: A Survey of Pediatricians. *Pediatrics, 138*(3). https://doi.org/10.1542/peds.2016-2127.

Jarrett, C., Wilson, R., O'Leary, M., Eckersberger, E., Larson, H. J., & Others. (2015). Strategies for Addressing Vaccine Hesitancy—A Systematic Review. *Vaccine, 33*(34), 4180–4190.

Johnston, J. M., & Sherman, R. A. (1993). Applying the Least Restrictive Alternative Principle to Treatment Decisions: A Legal and Behavioral Analysis. *The Behavior Analyst, 16*(1), 103–115.

Kahneman, D., Knetsch, J. L., & Thaler, R. H. (1991). Anomalies: The Endowment Effect, Loss Aversion, and Status Quo Bias. *The Journal of Economic Perspectives: A Journal of the American Economic Association, 5*(1), 193–206.

Krubiner, C. B., & Merritt, M. W. (2017). Which Strings Attached: Ethical Considerations for Selecting Appropriate Conditionalities in Conditional Cash Transfer Programmes. *Journal of Medical Ethics, 43*(3), 167–176.

Lagarde, M., Haines, A., & Palmer, N. (2007). Conditional Cash Transfers for Improving Uptake of Health Interventions in Low- and Middle-Income Countries: A Systematic Review. *JAMA: The Journal of the American Medical Association, 298*(16), 1900–1910.

Li, M., & Chapman, G. B. (2013). Nudge to Health: Harnessing Decision Research to Promote Health Behavior. *Social and Personality Psychology Compass, 7*(3), 187–198.

Lunze, K., & Paasche-Orlow, M. K. (2013). Financial Incentives for Healthy Behavior: Ethical Safeguards for Behavioral Economics. *American Journal of Preventive Medicine, 44*(6), 659–665.

Lynch, H. F. (2016). Introduction. In G. Cohen, H. F. Lynch, & C. Robertsonm (Eds.), *Nudging Health: Health Law and Behavioural Economics* (pp. 109–111). Baltimore: Johns Hopkins University Press.

Marteau, T. M., Ashcroft, R. E., & Oliver, A. (2009). Using Financial Incentives to Achieve Healthy Behaviour. *BMJ, 338*, b1415.

Miller, R. D. (1982). The Least Restrictive Alternative: Hidden Meanings and Agendas. *Community Mental Health Journal, 18*(1), 46–55.

Navin, M. C. (2017). The Ethics of Vaccination Nudges in Pediatric Practice. *HEC Forum: An Interdisciplinary Journal on Hospitals' Ethical and Legal Issues, 29*(1), 43–57.

Navin, M. C., & Largent, M. A. (2017). Improving Nonmedical Vaccine Exemption Policies: Three Case Studies. *Public Health Ethics, 10*(3), 225–234.

Novemsky, N., & Kahneman, D. (2005). The Boundaries of Loss Aversion. *JMR, Journal of Marketing Research, 42*(2), 119–128.

Nozick, R. (1969). Coercion. In S. Morgenbesser, P. Suppes, & M. White (Eds.), *Philosophy, Science, and Method: Essays in Honor of Ernest Nagel* (pp. 440–472). New York: St Martin's Press.

Nuffield Council on Bioethics. (2007). *Public Health: Ethical Issues.* Cambridge: Cambridge Publishers Ltd.

Nyhan, B., Reifler, J., Richey, S., & Freed, G. L. (2014). Effective Messages in Vaccine Promotion: A Randomized Trial. *Pediatrics, 133*(4), e835–e842.

O'Neill, O. (1991). Which Are the Offers You Can't Refuse? In R. Frey & C. Morris (Eds.), *Violence, Terrorism, and Justice* (pp. 170–195). Cambridge: Cambridge University Press.

Opel, D. J., Heritage, J., Taylor, J. A., Mangione-Smith, R., Salas, H. S., Devere, V., et al. (2013). The Architecture of Provider-Parent Vaccine Discussions at Health Supervision Visits. *Pediatrics, 132*(6), 1037–1046.

Pierik, R. (2016). Mandatory Vaccination: An Unqualified Defence. *Journal of Applied Philosophy.* https://doi.org/10.1111/japp.12215.

Ploug, T., & Holm, S. (2015). Doctors, Patients, and Nudging in the Clinical Context—Four Views on Nudging and Informed Consent. *The American Journal of Bioethics: AJOB, 15*(10), 28–38.

Ranganathan, M., & Lagarde, M. (2012). Promoting Healthy Behaviours and Improving Health Outcomes in Low and Middle Income Countries: A Review of the Impact of Conditional Cash Transfer Programmes. *Preventive Medicine, 55*(Suppl), S95–S105.

Rawls, J. (1971). *A Theory of Justice: Revised 2009 Edition.* Cambridge, MA: Harvard University Press.

Rossi, J., & Yudell, M. (2012). The Use of Persuasion in Public Health Communication: An Ethical Critique. *Public Health Ethics, 5*(2), 192–205.

Saghai, Y. (2013). Salvaging the Concept of Nudge. *Journal of Medical Ethics, 39*(8), 487–493.

Saghai, Y. (2014). Radically Questioning the Principle of the Least Restrictive Alternative: A Reply to NirEyal': Comment on "Nudging by Shaming, Shaming by Nudging". *International Journal of Health Policy and Management, 3,* 349–350.

Salmon, D. A., Moulton, L. H., Omer, S. B., DeHart, M. P., Stokley, S., & Halsey, N. A. (2005). Factors Associated with Refusal of Childhood Vaccines Among Parents of School-Aged Children: A Case-Control Study. *Archives of Pediatrics & Adolescent Medicine, 159*(5), 470–476.

Thaler, R., & Sunstein, C. (2008/2009). *Nudge. Improving Decisions About Health, Wealth, and Happiness.* London: Penguin.

Verweij, M., & Dawson, A. (2004). Ethical Principles for Collective Immunisation Programmes. *Vaccine, 22*(23–24), 3122–3126.

Voigt, K. (2017). Too Poor to Say No? Health Incentives for Disadvantaged Populations. *Journal of Medical Ethics, 43*(3), 162–166.

Wang, E., Clymer, J., Davis-Hayes, C., & Buttenheim, A. (2014). Nonmedical Exemptions from School Immunization Requirements: A Systematic Review. *American Journal of Public Health, 104*(11), e62–e84.

Warwick, D. P., & Kelman, H. C. (1973). Ethical Issues in Social Intervention. In G. Zaltman (Ed.), *Processes and Phenomena of Social Change*. Hoboken: Wiley.

Wertheimer, A. (1989, April). Review of Coercion (1987), by R. Nozick. *Ethics, 99*, 642.

WHO. (2014). Report of the SAGE Working Group on Vaccine Hesitancy. Retrieved April 2018, from http://www.who.int/immunization/sage/meet-ings/2014/october/1_Report_WORKING_GROUP_vaccine_hesitancy_final.pdf

Wigham, S., Ternent, L., Bryant, A., Robalino, S., Sniehotta, F. F., & Adams, J. (2014). Parental Financial Incentives for Increasing Preschool Vaccination Uptake: Systematic Review. *Pediatrics, 134*(4), e1117–e1128.

Yamin, D., & Gavious, A. (2013). Incentives' Effect in Influenza Vaccination Policy. *Management Science, 59*(12), 2667–2686.

Yaqub, O., Castle-Clarke, S., Sevdalis, N., & Chataway, J. (2014). Attitudes to Vaccination: A Critical Review. *Social Science & Medicine, 112*, 1–11.

Zamir, E. (2015). *Law, Psychology, and Morality: The Role of Loss Aversion*. Oxford: Oxford University Press.

CHAPTER 4

Fairness, Compulsory Vaccination, and Conscientious Objection

Abstract This chapter presents an argument for compulsory vaccination and against allowing non-medical vaccine exemptions. The argument is based on the idea that the proper aim of vaccination policies should be not only herd immunity but also a fair distribution of the burdens entailed by its realization. I argue that a fairness requirement need not and should not be constrained by a principle of liberty and a principle of least restrictive alternative. Indeed, I argue how compulsory vaccination is more successful than other types of vaccination policies at satisfying the principles of fairness, least restrictive alternative, and maximizing expected utility, once these principles have been properly understood.

Keywords Fairness • Compulsion • Compulsory vaccination • Conscientious objection

IMPLICATIONS OF THE PRINCIPLE OF LEAST RESTRICTIVE ALTERNATIVE

In the previous chapter, I have suggested that if we assume that the realization of herd immunity is the goal of vaccination policies, then according to the principle of least restrictive alternative (PLRA), there are ethical reasons to implement the least restrictive policy capable of achieving that

© The Author(s) 2019
A. Giubilini, *The Ethics of Vaccination*, Palgrave Studies in Ethics and Public Policy,
https://doi.org/10.1007/978-3-030-02068-2_4

goal, where restrictiveness is measured in terms of degree of limitation of individual autonomy or liberty. In this chapter, I am going to question that assumption.

Before doing that, let me first emphasize two implications of the discussion in the last chapter. First, as we saw, within that perspective, compulsory vaccination should be seen as a measure of last resort: depending on the socio-economic and cultural context, some of the less restrictive options discussed in Chap. 3 could be sufficient for the realization of herd immunity. Convincing people to vaccinate without coercing them is always preferable, and where some coercion is necessary to achieve herd immunity, a lower degree of coercion is always preferable. Besides, coercive policies would be necessary only where outright vaccine refusal is a significant factor in low vaccination rates; where this is not the case, alternative and less restrictive policies such as nudging, improving accessibility and minimizing logistic barriers to vaccination are not only ethically preferable on the basis of the PLRA but probably also very effective (Beard et al. 2017).

Second, in virtue of the same PLRA, in case of certain coercive vaccination policies such as mandatory vaccination, it makes good ethical sense to grant non-medical exemptions or "conscientious objection" to people who are opposed to vaccines. Since one of the parameters for measuring restrictiveness is the amount of people who are burdened with restrictions on liberty or autonomy (the other being the magnitude of autonomy restriction for the worse off), it seems we should grant non-medical exemptions to a certain number of people who have deeply held beliefs against vaccination, provided they are not too large a group. A policy that restricts the liberty of fewer people is preferable to one that restricts the liberty of more people. If conscientious objectors constitute, say, 2% of the population in a given jurisdiction, a PLRA would demand that these people be exempted from vaccination mandates, provided that the number of those exempted for medical or age-related reasons is also sufficiently low to guarantee that the total number of exemptions does not compromise herd immunity.

As is the case with many other instances of conscientious objection, particularly in the medical context (e.g., doctors' conscientious objection to performing abortions), one problem is that, once conscientious objection is allowed, it is difficult to keep exemption rates sufficiently low without introducing some discrimination between those who are granted exemptions and those who are not (Giubilini 2014). Moreover, as is always the case whenever the obscure notion of "conscience" (Giubilini 2016) is brought up, many would appeal to conscience for mere reasons of convenience, such as the desire to free-ride on herd immunity. Thus, for instance, in the US, where child vaccination is a condition for enrolling children in day care or state schools, the vast majority of states allow conscientious objection, and almost all these states saw a twofold increase of exemption rates from the 2005–06 to the 2012–13 school year. In some cases, such increase threatened herd immunity; for example, exemption rates in Oregon increased from 3.4% to 6.4% over the same period (Wang et al. 2014). However, even where conscientious objection is normally granted, it might be possible to achieve herd immunity by making the exemption procedure particularly burdensome—for example, by requiring objecting parents to attend vaccine information sessions or to go through certain bureaucratic procedures—thus discouraging as many people as possible from applying for exemptions (Salmon and Siegel 2001; Bester 2015; Navin and Largent 2017). As Mark Navin and Mark Largent (2017) have argued, burdensome exemption procedures are likely to be the most ethical vaccination policies, since they promote both (parental) freedom not to vaccinate and effectiveness in boosting vaccination rates above the herd immunity threshold. Indeed, if and where such policies are actually effective, Navin and Largent seem to have a point. I will say more about conscientious objection and burdensome exemption procedures in the conclusion of the chapter.

These two implications of the PLRA—namely, that compulsory vaccination is a measure of last resort and that a certain number of conscientious objections to vaccination should be tolerated—are consistent with the idea that there can be *pragmatic* reasons for aiming at *universal* vaccination coverage, which is very likely to require unqualified compulsory vaccination. By "unqualified compulsory vaccination" I mean a policy that makes non-vaccination illegal for anyone for whom vaccination is not medically contraindicated, without any accommodation for conscientious objection. Since there will always be, in any given population, a certain percentage of people who cannot be vaccinated for medical reasons and who will contribute to lowering the vaccine coverage rate, realization of herd immunity might still require, in practice if not in principle, that *all* those who can be vaccinated without significant medical contraindication be vaccinated (Dawson 2007, p. 173). However, while these pragmatic considerations are certainly important when deciding which vaccination policy to implement, they are merely contingent and not principled reasons for unqualified compulsory vaccination: if the circumstances were such that we could be sure that enough people were vaccinated anyway, such pragmatic considerations would no longer apply. Since gathering data about vaccine coverage rate is routinely done by governments, it would not be very difficult to verify at any time whether aiming at universal vaccination coverage is necessary to at least realize herd immunity. Actually, the PLRA would demand to regularly verify such contingencies in order to implement, at any given time, the vaccination policy that is least restrictive of individual autonomy: according to the PLRA, there is an ethical imperative not to aim at universal vaccination coverage in order to realize herd immunity through coercive measures if *less* people can be coerced into vaccinating without threatening herd immunity. This requirement implies that we should switch from a more coercive to a less coercive policy when the more coercive policy is no longer necessary for the realization of herd immunity.

However, what I want to investigate here is whether there are *principled*, that is, ethical, reasons for unqualified compulsory vaccination. If they existed, such reasons would apply regardless of whether less restrictive measures for realizing herd immunity were available. This means that they would render irrelevant either the PLRA, for example, by claiming that, other things being equal, more restrictive policies can be justified, or the idea that herd immunity should be the goal of vaccination policies, for example, by claiming that vaccination policies should have more demanding targets. Here, I will pursue the second strategy: I will argue that vaccination policies ought to aim at *universal* vaccination coverage, rather than merely at herd immunity, and that therefore they ought to take the form of unqualified compulsion. This claim is consistent with the existence of the PLRA: I am not denying that less restrictive policies are ethically preferable to more restrictive ones, *other things being equal*. The point is rather that, so to speak, other things are *not* equal: as I shall argue, we ought not to choose the policy that is less restrictive among those that can achieve herd immunity, but we ought to choose the policy that is less restrictive *and* that would achieve universal vaccination coverage (where "universal" does not include, of course, individuals who cannot be vaccinated for medical reasons or for age limits). What is not equal is the appropriate aim of vaccination polices: not herd immunity—which is what was equal in the list of policies discussed in the previous chapter—but a fair distribution of the burdens entailed by herd immunity. That vaccination policies "ought to" aim at universal vaccination coverage means that the reasons for aiming at universal vaccination coverage are not merely pragmatic, but indeed ethical: there is an ethical principle that demands aiming for universal vaccination.

In this chapter, I will argue that there are strong ethical reasons for enforcing compulsory vaccination, rather than less coercive vaccination policies, and for not allowing non-medical vaccination exemptions, and hence for enforcing unqualified compulsory vaccination. Of course, one might object that such ethical reasons would be worthless if it turned out that compulsory vaccination is less effective than other less coercive measures, that is, that vaccination coverage rates would (likely) be lower where compulsory vaccination is implemented. For example, one might suggest that compulsory vaccination could backfire by reinforcing people's anti-vaccination sentiments and by encouraging them to find ways to escape the requirement. After all, one might observe, given the importance of a good like herd immunity, we should prioritize effectiveness over other ethical values; or, put more sharply, a solution cannot be truly ethical if it prevents such an important public good as herd immunity from being realized. However, as I will argue below, even if it were true that compulsion would discourage people from vaccinating, compulsory vaccination would not backfire if adequately implemented.

THE ETHICAL CASE FOR COMPULSORY VACCINATION

Claims to the effect that vaccination should be compulsory are typically grounded in considerations about the harm, or risk of harm, that non-vaccination presents to other people (e.g., Flanigan 2014; Bambery et al. 2013). Alternatively, they might appeal to the so-called clean hand principle (Brennan 2018), which morally prohibits people from becoming accomplices in the collective harm that would result from the failure to realize herd immunity. On yet other views, compulsory vaccination is justified by the fact that the state has an obligation "to guard the common good of herd immunity" in order to protect vulnerable people (Pierik 2016, p. 7). If you agree with my analysis of the relationship between individual and collective responsibility from Chap. 2, you will see that all of these justifications are problematic. As I have suggested, realization of herd immunity and therefore prevention of harm are collective enterprises: no individual can achieve these goals or make a significant contribution towards them. Since each individual contribution to the collective harm produced by low vaccination rates is negligible, and since each individual contribution to the "common good" (Pierik 2016) of herd immunity (which is more appropriately described as a "public good") is negligible, it is difficult to argue that *each individual* should be subject to a legal obligation to be vaccinated, as compulsory vaccination would require—at least if we accept the validity of the PLRA (which, as I have said, I have no intention to reject here). In fact, those who provide arguments for compulsory vaccination typically do not reject the PLRA as applied to policies aimed at minimizing risks, and indeed some of them explicitly endorse it. For instance, both Flanigan (2014, p. 20) and Pierik (2016, p. 8) state that vaccination policies aimed at protecting vulnerable individuals from risks of harm should be informed by the PLRA. Thus, their claims result much more attenuated than their initial intentions might suggest: ultimately, these arguments imply that vaccination should be compulsory only when there is some significant risk of harm posed by non-vaccination. In other words: an argument for compulsory vaccination would be very easily defeated by appealing to the PLRA as long as there are less coercive policies that could guarantee herd immunity. So considerations about harm to others cannot provide a principled justification for compulsory vaccination if we accept the PLRA, as I do here and as those who argue for coercive vaccination policies normally do.

According to Jason Brennan, the clean hand principle makes it impermissible even for libertarians to appeal to individual freedom against state-enforced vaccination, because in this case individual freedom would make one morally complicit in collective harm. But the clean hand principle is subject to the same objection about the relationship between collective and individual obligations: you cannot justify an individual obligation, a legal one in this case, to be vaccinated simply by appealing to the notion of collective harm. Even if you think, as a matter of private morality, that you should keep your hands clean and not make any contribution to some collective harm, you would need a more fundamental ethical principle to justify the *enforcement*, by the state, of a clean hand principle in case of insignificant individual contribution to collective harm. According to Brennan, the clean hand principle in the context of vaccination is an *enforceable* moral principle within a libertarian framework because it "stops individuals from causing harm" (Brennan 2018, p. 40). But this does not seem to be right: where there is a sufficiently high vaccination rate, getting my hands dirty—that is, refusing vaccination—does not cause any harm; and more generally, whether my hands are clean or dirty does not make a difference to whether herd immunity is realized. So why should I be required to keep my hands clean and contribute to the public good of herd immunity, if my contribution makes only a very small or even negligible difference to herd immunity? What is the political value or the impact on others of the mere fact that my hands are clean? Notice that I am not arguing here against Brennan's claim that vaccination should be compulsory, nor am I interested in how a libertarian could justify compulsory vaccination while remaining libertarian. Actually, I might even agree with Brennan's claim that "the libertarian must endorse *something like* the clean hands principle, and, further, must accept in some cases it is permissible to use coercion against the individuals who constitute the collective performing the rights violation or causing the harm" (Brennan 2018, p. 42). Whether I would be prepared to endorse this claim depends on how we interpret the phrase "*something like* the clean hands principle". This phrasing is quite vague. What does it mean? Something like, but not exactly, the clean hands principle? If so, I agree with this quite vague formulation, although it is not a very informative one; my point is simply that the "clean hand" principle itself (not "something like" it) does not do the work Brennan wants it to do.

One way to support the clean hand principle is simply dogmatic: keeping one's hands clean and not being moral accomplices of bad outcomes is good in and of itself, full stop. This is, for example, the approach that grounds the Catholic doctrine of the moral wrongness of formal cooperation in wrongdoing. Another possible approach could be considered deontological: getting one's hands dirty would violate some established ethical code. And there might be other ethical theories that can justify a moral obligation not to be accomplices in collective harms even when the individual contribution is negligible. Examples include contractualism (see, e.g., Giubilini et al. 2018), that is, roughly, the theory according to which we should act on the basis of principles that other people could not reasonably reject, or rule consequentialism, that is, the theory according to which everybody should follow those simple rules that, if followed by everyone else, would produce the best consequences, such as "be vaccinated". However, even if we want to accept any of these justifications (dogmatic, deontological, consequentialist, or contractualist) for the clean hand principle, they can hardly be taken to warrant the *legal* enforceability of a moral principle: some of them only concern the sphere of individual morality, and those that aim at informing policymaking (such as contractualism and perhaps some versions of rule consequentialism) require to commit to comprehensive moral theories that many people would not accept. In order to enforce a moral principle, the principle should ideally be one of those uncontroversially taken to have some relevance for policymaking, no matter what comprehensive moral theory one embraces.

As it happens, there is one such ethical principle that does justify the legal enforceability of the clean hand principle and, more generally, that justifies an individual obligation—either moral or legal—to be vaccinated, starting from a collective obligation to realize herd immunity. This is, as I have argued in Chap. 2, a principle of fairness. For one thing, fairness is a principle that is typically taken to have relevance for policymaking: normally, we want public policies to be fair (the precise definition of which I will elaborate on later). Besides, unlike comprehensive moral theories such as contractualism or rule consequentialism, a principle of fairness is certainly widely shared and in line with commonly held moral intuitions: it is a principle most reasonable people would subscribe to when it comes to distribution of certain burdens, no matter what their ethical approach is. As such, it clearly qualifies as a reasonable principle—on any plausible understanding of "reasonable"—around which to design vaccination policies. Of course, the problem would be that of balancing it with other principles that reasonable people consider important as grounds for policymaking, such as individual liberty and maximization of expected utility (see, e.g., Selgelid 2009). As I will suggest in this chapter, even in light of this balancing, there are reasons to give fairness in the distribution of burdens priority over individual liberties in the formulation of vaccination polices and therefore to subject everyone to a legal obligation to be vaccinated in spite of the infringement on individual liberty.

John Stuart Mill is one of the champions of liberalism; he famously grounded the justification for coercive state interventions on the so-called harm principle—roughly, the principle that the only purpose for which states may and ought to use its coercive power is to protect individuals from liberty restrictions or harm caused by other individuals' behaviour. But even Mill at some point had to introduce considerations of fairness to justify the legal enforceability of individual moral obligations in the context of prevention of *collective* harm. Even in his view, appeal to harm prevention was not sufficient to ground certain individual moral and legal obligations in those cases where some individual contribution to a collective harm—or to the failure to realize a collective benefit—would be negligible or in any case would not determine whether other people would be harmed. Instead, he formulated the following principle, which clearly refers to fairness requirements:

> everyone who receives the protection of society owes a return for the benefit, and the fact of living in society renders it indispensable that each should be found to observe a certain line of conduct towards the rest. This conduct

consists (…) in each person's bearing his share (…) of the labours and sacrifices incurred for defending the society or its members from injury and molestation. (Mill 1859, pp. 140–141)

This principle could ground a legal obligation for each citizen to vaccinate themselves or their children, as an instance of a more general obligation to fairly contribute to important public goods (as people normally do, for example, through taxation.). Granted, the notion of fairness involved in Mill's formulation is different from the notion of fairness I have defended in Chap. 2: while Mill seems to appeal to a principle of reciprocity and to a duty not to free-ride, my argument in Chap. 2 referred to fairness in the distribution of the burdens entailed by the collective obligation to realize herd immunity. But the important point I want to emphasize here is that, even within Mill's perspective, a mere reference to prevention of harm to others is not sufficient to justify individual moral or legal obligations.

Before I proceed, it is worth clarifying that there are different ways of understanding "fairness". Perhaps the most fundamental distinction is between fairness as *equity* and fairness as *equality*. The former implies that a fair distribution of burdens is one where everyone is burdened according to some morally relevant criterion, such as her capacity to bear the burden, or considerations of desert or lack thereof. The latter implies that a fair distribution is one where everyone is burdened *the same*, regardless of capacities or of any other factor. In Chap. 2 I assumed that a fair distribution of the burdens of the collective responsibility to realize herd immunity was one where everybody without medical contraindications against vaccination be vaccinated. In light of the distinction I have just drawn, this represents an equitable, not an equal distribution of the burdens: those who do not have the capacity to bear the burden of vaccination are not subject to the fairness-based obligation to be vaccinated. In general, the same principle of fairness as equity applies when it comes to policymaking regulating any distribution of burdens. Fairness as equity, for example, is the reason why taxation as a form of contribution to public goods is often progressive, rather than proportional or flat: people are not subject to the same tax rate, but are taxed in consideration of their capacity to contribute to state expenditures, for instance, on the basis of factors such as income, number of dependants, number of houses owned, and so on. The same criterion should be applied—or so I have argued—to the case of vaccination as a contribution to herd immunity, except that, in the case of vaccination, there are only two options: either one contributes by being vaccinated, or one does not contribute by being exempted for medical reasons, that is, because one does not have the capacity to contribute in the same sense in which others have such capacity.

Thus, fairness is an important *ethical* and *social* value when it comes to sharing burdens required by the preservation of public goods. In the remainder of this section, I will explain in what sense fairness is an important *ethical* value; this explanation will clarify why fairness is not in conflict with the PLRA, when the PLRA is properly understood in the context of vaccination policies. In the next section, I will explain in what sense fairness is an important *social* value; this explanation will clarify why fairness is not in conflict with a principle of maximization of expected utility (which might seem obvious, but which some people have questioned).

Now, as for its *ethical* aspect, I have said that there are reasons to prioritize fairness in the distribution of burdens over individual liberty in the formulation of vaccination policies. In other words, from an ethical point of view, fairness trumps liberty in the context of vaccination policies. Admittedly, the claim is controversial: Nozickian libertarians, for example, would claim that bodily autonomy is a fundamental liberty, and therefore the state is not justified in violating bodily autonomy to promote a good that is valuable for the community (for a discussion, see Navin 2015, p. 182). However, one might argue that there are reasons to reject this implication within a libertarian framework even if, as I have suggested above, we have reasons to reject Brennan's appeal to the "clean hand" principle. As suggested by Navin, within a libertarian framework, the state is still justified in implementing coercive policies that infringe upon certain individual rights if such policies are necessary to prevent harm to others, because harm to others—for example, in the form of infectious diseases—does limit their liberty (Navin 2015, p. 182). But Navin's argument is problematic: if my non-vaccination does not significantly increase the risk posed on others, especially when vaccination coverage is very high, how can libertarians consistently justify coercive vaccination policy, given that my non-vaccination would not significantly threaten other people's liberty? The answer is that they cannot without also accepting that, in the case of collective harm, the harm principle does come with a fairness principle attached, as a matter of basic ethical requirement. Some libertarians might not accept that fairness plays this important role in policymaking—though rejecting a fairness requirement would make it difficult to accept "something like" the clean hand principle. But their refusal to accept the fairness requirement in policymaking tells more about the low ethical standard libertarians apply to policymaking than about the importance of the fairness requirement in the distribution of burdens entailed by public goods.

Moreover, it might be argued, even libertarians would concur that a state should ensure that individuals make a *fair* contribution to the prevention of such harm. Mark Navin, for example, has defended a principle of fair distribution of certain burdens from within a libertarian perspective—although he is not a libertarian—by way of analogy with other cases where he thinks at least certain libertarians do endorse such principles of fairness. For example, Navin argues, libertarians would agree that the state is justified in ensuring that everybody pays their fair share of taxes to support military and police expenditures, in order to protect national security and public safety, which are necessary for protecting individual liberty. And as we have said above, fairness is also what ultimately justifies the "clean hand" principle that some libertarians want to incorporate within their ethical perspective to justify coercive vaccination policies. Thus, even within a libertarian framework, it can be argued that fairness does play a fundamental role in determining how the burdens entailed by the preservation of certain goods and prevention of certain harms should be distributed (Navin 2015, p. 182).

So far, I have established, at the very least, that fairness in the distribution of the burdens entailed by the preservation of public goods (or, in a libertarian framework, of those goods that prevent serious harm to individuals and thus compromise their liberty) is an important value in the formulation of public policies, and of vaccination policies in particular. However, I want to argue for something stronger: my claim is that fairness is a value that need not and should not be compromised by being balanced against other values involved in policymaking, such as individual liberty and expected utility (i.e., the realization of herd immunity).

With regard to the three core ethical values informing reasonable public policies, it would seem that the relevant question is: what is the relative importance of fairness, expected utility, and liberty in the formulation of public policies aimed at realizing herd immunity? More precisely, if this is the relevant kind of question, it would seem that at this stage of my argumentation the actual question should be restricted to the relative weight of fairness and the PLRA and to the relative weight of fairness and expected utility. After all, we already know how to balance liberty and expected utility: we have already established that one of the goals of vaccination policies is the realization of herd immunity, and that balancing the value of expected utility against the value of individual liberty leads to the reformulation of a principle of liberty in terms of the PLRA: we ought to implement the least restrictive policy that is effective in realizing herd immunity. So what remains to be done is to establish how to balance fairness against the PLRA and fairness against expected utility. The relevant question seems, then, to be the following: provided that we ought to realize herd immunity, how should we weigh fairness in the distribution of burdens entailed by the realization of herd immunity against the PLRA and fairness against expected utility?

I want to claim is that this is not the right question to ask. The more fundamental, and more important, questions are rather the following: is fairness *really* in conflict with the PLRA, understood as a principle protecting individual liberty? Or, is fairness *really* in conflict with expected utility, understood as the realization of herd immunity, in the formulation and implementation of vaccination policies? If the answer to both questions were negative, then the questions about the relative importance of fairness and the PLRA and about the relative importance of fairness and expected utility would not arise at all, because we would not need to balance them one *against* the other. And indeed, the reason why I claim that fairness "need not" and "should not" be balanced against other values is that there is no actual conflict between fairness and the PLRA and between fairness and expected utility.

Now, I am of course aware that, at a first glance, it would seem that fairness and the PLRA are in conflict with one another, and therefore that if compulsory vaccination were enforced on the model of compulsory taxation to ensure that everybody made their fair contribution to the public good, the PLRA would automatically be violated: after all, herd immunity could be realized even with policies that would allow some people to obtain non-medical exemptions or that would simply nudge or incentivize people to vaccinate. Therefore, the PLRA, to the extent that it requires implementing non-compulsory policies whenever non-compulsory policies are consistent with the realization of herd immunity, means that fairness is violated in principle—although it might still be realized in practice in the extremely unlikely event that *everyone* decided to be vaccinated or to vaccinate their children even when they are not be compelled to do so. Even Mark Navin, one of the advocates of the importance of fairness in the current debate on vaccination policies, thinks that fairness needs to be balanced against something like the PLRA when it comes to vaccination policies, and therefore that, in spite of the fairness requirement, compulsory vaccination should not be enforced when less restrictive policies are effective in realizing herd immunity. As Navin writes, "the state has the authority to coerce vaccination, though there are good reasons for it to use as little coercion as is necessary to achieve the goal of herd immunity" (Navin 2015, p. 12); elsewhere, as we will see below, he argues that non-medical exemptions to vaccination mandates should be allowed as long as herd immunity is realized, in order to preserve liberty to the greatest extent possible (Navin and Largent 2017).

However, the conflict between fairness and the PLRA is only apparent. The analogy between vaccination policies and taxation policies is a very good one, because it shows that when we need to ensure that important public goods are preserved or realized, as is the case with taxation, the appropriate goal to pursue is not only the realization or preservation of such public goods but also that everybody makes their fair contribution to them. As we have seen in Chap. 1, (1) everybody benefits from public goods and from herd immunity in particular, and, as we have seen in Chap. 2, (2) we do have a collective obligation to realize herd immunity, and the vast majority of us has the capacity to make our contribution to the fulfilment of this collective obligation at a small personal cost. Therefore, it would be unethical for a policy to aim at the realization of herd immunity without considering the fairness implications of (1) and (2): everybody ought to contribute to herd immunity (1) as a matter of reciprocity and avoid free-riding and (2) as a matter of fair distribution of burdens. To compare, people are normally not exempted from paying their share of taxes just because they do not ethically approve of some ways in which their government spends public money, such as in the case of pacifists, or because they are afraid that it would be unsafe to spend public money in certain ways, for example, in case someone thought that some military operations would trigger reactions from terrorists in one's own country. These exemptions would simply be unfair, given that, for instance, even pacifists benefit from the public good of national security preserved through military forces (reciprocity and free-riding consideration) and that also pacifists are able to contribute to the fulfilment of the collective obligation to preserve national security (consideration of the distribution of burdens).[1] Thus, fairness is not simply

---

[1] I am assuming here that there is such a collective obligation; if you think that there is not, you can replace the national security example with the example of any other important public good that you think we have a collective obligation to preserve, such as clean air or a publicly funded health system.

one of the principles informing public policies aimed at regulating the distribution of burdens entailed by realization or maintenance of certain public goods but actually one of the goals of such policies. Fairness and the PLRA are not in conflict with one another because in the context of policies aimed at preserving important public goods they are not on the same level of importance, and actually the attempt to balance them against one another presupposes a categorical mistake: fairness is one of the ends of vaccination policies, while the PLRA is a way of determining which means are appropriate to the ends. That fairness is "one of the ends" of vaccination policies does not mean, of course, that fairness is as important a goal as other goals of vaccination policies and in particular the protection of population's health; but it does mean that it is something vaccination policies should *aim* at in pursuing their more fundamental goal. We can think of fairness as a *subordinate* or *secondary* goal of vaccination policies: we do not enforce vaccination policies *in order* to promote fairness, but once we decide to enforce vaccination policies in order to realize herd immunity and prevent harm, fairness does become one of the goals of these policies, because herd immunity should be realized *fairly*. What constitutes the "least restrictive alternative" depends on the ends that a policy is supposed to promote (for instance, the least restrictive policy that is able to realize herd immunity), but fairness is itself one of the ends that a policy aimed at protecting or realizing an important public good should promote, together with its more fundamental goal of realizing the public good in question. In other words, a successful vaccination policy is not only one that achieves herd immunity but one that achieves herd immunity *fairly*. Of course, once we have established that the goal is a fair realization of herd immunity, we still need to apply the PLRA and try to achieve the goal through the least restrictive policy possible; but we need not and indeed should not reach a compromise between the ends and the ethical restrictions about which means should be used to achieve those ends. Doing that would undermine the very purpose of the policy. Unfortunately, in practice, the least restrictive policy that would allow to fairly realize or preserve herd immunity is very likely to be unqualified compulsory vaccination, in the same way as unqualified compulsory taxation is very likely to be the least restrictive policy that can ensure that *everybody* makes their fair economic contribution to certain public goods. Many people would not vaccinate and would not pay their taxes if they were not compelled. True, compulsory vaccination also happens to be a very restrictive policy, but this does not mean that it violates the PLRA; it only means that, given the legitimate ends of vaccination policy, the PLRA has limited scope for influencing the level of restrictiveness of the policy.

One might observe at this point that, even with compulsion, some people would still be able to get away with refusing vaccination in the same way as some people manage to evade taxes and go unpunished. Therefore, my argument seems to involve a *reductio ad absurdum* because it seems to entail that, in order to fulfil the requirements of both fairness and expected utility, even more restrictive policies would be required, such as forced vaccination—for example, with health and if necessary police authorities visiting individual households and performing the inoculation. But clearly, forced vaccination is unacceptable, or at least so the objection might go (and I shall assume, for the sake of argument, that this is true). Thus, since my argument has an unacceptable implication, some of its premises (e.g., about the importance of fairness) or something in its logical steps (e.g., the analogy with taxation), or both, must be wrong. However, I am not arguing for forced vaccination here, and I do not think that my argument would commit me to endorsing forced vaccination. Therefore, I do not think that the argument leads to a *reductio ad absurdum*.

Why do I say that my argument does not imply a defence of forced vaccination? The reason is not, as many might be thinking, that forced vaccination would violate a principle of bodily integrity or bodily autonomy and therefore that such principles would morally outweigh the moral force of fairness when it comes to forced vaccination. The principles of bodily integrity or bodily autonomy need to be properly understood and, I would add, properly downplayed.

For one thing, bodily autonomy and bodily integrity are only prima facie principles that can permissibly be violated, and indeed are often violated, when something more valuable has to be promoted, and effectiveness and fairness in vaccination policies are very important values. Thus, for example, sometimes forced medical treatments are carried out in spite of the violation of bodily integrity and bodily autonomy—such as in the case of people with certain mental disorders or in the case of chemical castration for some sex offenders in certain states, such as currently California or Indonesia; and sometimes other equally restrictive measures are forced upon individuals to protect public health—such as in the case of quarantine or isolation to prevent or contain epidemics. For another thing, we should not and normally do not attribute such a great importance to the principle of bodily integrity or bodily autonomy; for example, most of us are fine with fluoride being added to running water, often without people being adequately notified about it, for the sake of individual and public health (specifically, preventing tooth decay). However, people who are opposed to vaccines and who appeal to principles like bodily integrity and bodily autonomy are unlikely to accept the hypothetical administration of vaccines through the same means, for example, if, as a thought experiment, vaccines could simply be added to running water. This suggests that the real (ethical) concern is not about violation of bodily integrity or autonomy, but specifically about vaccines.

Thus, the reason why my argument does not lead to a defence of forced vaccination has nothing to do with a principle of bodily integrity or bodily autonomy—and after all, even compulsory vaccination would be inconsistent with a principle of bodily integrity or autonomy. Rather, I "only" argue for compulsory vaccination because I am interested in the ethical principles involved in vaccination policies and their relative priority: expected utility, fairness, and PLRA. An adequately implemented compulsory vaccination policy would meet all the requirements set by these principles, without having to implement even more restrictive policies.

Thus, in an ideal world where everyone abides by the law, compulsory vaccination satisfies all the ethical principles of policymaking. True, in the real world many people do not observe the law and many of them get away with it. Therefore, likely, no compulsory policy would achieve a 100% observance, thus falling short of meeting the requirements of fairness in the distribution of burdens. If someone thinks that this empirical consideration justifies the implementation of forced vaccination, I would not think that their point of view is *obviously* unacceptable. The issue at stake is whether in-principle considerations should be replaced by more down-to-earth practical considerations when formulating public policies; but this is an issue I am happy to leave open here. I will only note that if we had to replace *in-principle* considerations with more practical considerations, then we might as well sacrifice the principle of fairness and implement less restrictive policies that would allow a few people to unfairly free-ride on herd immunity as long as herd immunity is not threatened, and this approach would *a fortiori* rule out any justification for forced vaccination. But I think that a public policy should take important ethical principles into account.

So far, I have argued that fairness is not in conflict with the PLRA because the two principles are not on the same level, and therefore that compulsory vaccination is ethically justified because it promotes fairness without violating the PLRA. What about the second possible conflict between the values at stake in vaccination policies, namely, that between fairness and expected utility? I will address this question in the next section.

THE SOCIAL RELEVANCE OF FAIRNESS

The second aspect of fairness that I have mentioned above, namely, its *social* relevance, is what explains why fairness and maximization of expected utility are not in conflict with each other. Fairness in the distribution of burdens entailed by public goods has social relevance in the sense that it can influence people's behaviour in such a way as to determine whether or not socially important public goods become realized. Not only, as we have seen, maximizing expected utility by ensuring that as many people as possible are vaccinated promotes fairness; also, promoting fairness in the distribution of burdens entailed by the fulfilment of collective obligations is likely to contribute to the fulfilment of the collective obligation. How so? Because fairness is also likely to be *instrumentally* valuable: knowing that the burdens to which they are subjected are fairly distributed among people around them provides individuals with additional motivation to choose to bear their share of the burden, or at least removes some psychological barriers to cooperation. In other words, ensuring a fair distribution of the burdens of herd immunity would solve the so-called problem of assurance (Navin 2015, pp. 179–180), which Rawls identified as deriving from the fact that people would presumably be willing to contribute their fair share to a public good only if they knew that others are doing the same (Rawls 1971/1999, p. 236). Therefore, aiming at a fair distribution of burdens makes it more likely that the objective in question—in our case the realization of herd immunity—is achieved. The instrumental value of fairness might partially address the concern discussed above that even with compulsory vaccination some people would still be able to get away with non-vaccination: a fair distribution of the burdens of vaccination would ensure that free-riders are not as plentiful as they would otherwise be.

These claims about the social relevance of fairness are supported by a body of psychological evidence and evidence from neuroscience, including functional magnetic resonance imaging (fMRI), suggesting that being treated fairly is associated with positive emotions that motivate people to cooperate in collective enterprises, while unfairness produces the exact opposite consequences (see, e.g., Tabibnia et al. 2008). Cooperation is understood in the context of such psychological studies as "doing one's share to maximize public goods rather than working individually to maximize personal goods" (Tabibnia and Lieberman 2007, p. 91), a definition which fits well with what I have argued is an individual moral obligation in the case of vaccination (except that in many cases vaccination would also maximize an individual's personal good). Fairness in the distribution of rewards, burdens, and obligations is a powerful motivator to act cooperatively towards common goods (Tabibnia and Lieberman 2007): we know that humans tend to contribute to common objectives when they know that there are fair arrangements in place, but, despite the desirability of the common objective, they tend not to make their contribution under unfair arrangements. For instance, experiments in the context of the "Ultimatum Game" have shown that people are less likely to collaborate in order to achieve a common good if they perceive they have been treated unfairly. In the standard Ultimatum Game, one player, the proposer, is asked to split a sum of money with another player, the responder; the proposer can decide in what proportion (e.g., 80/20, or 50/50, etc.) to split a sum of money, and the responder can decide whether to accept or reject the offer; if the responder rejects the offer, neither player gets anything, whereas if the responder accepts, each one gets the sum that has been decided by the proposer. In spite of the financial gain in accepting an unfair offer, a perceived unfair offer of 20% of the sum—as opposed to a perceived fair offer represented by an equal share—has a 50% chance of being rejected by the responder, even if this means her losing out on that 20%; simply, the emotional reaction of the responder is to punish the proposer for the unfair treatment even at a personal and at a collective cost (Guth et al. 1982; Camerer and Thaler 1995). In other words, people fail to collaborate towards a common good—in this case, represented by retaining the sum of money, even if unfairly split—if they perceive others are not acting fairly towards them. In fact, studies have shown "neural overlap between the pleasure of being treated fairly and that associated with the intention to cooperate", as well as a decrease of such brain activity in cases of unreciprocated cooperation (Tabibnia and Lieberman 2007, p. 96). It is not unreasonable, and indeed it seems to me quite plausible, to suppose that the same psychological mechanisms are triggered when fairness issues arise in the context of individual contributions to herd immunity.

Thus, these studies suggest that fairness does contribute towards the maximization of expected utility by solving the problem of assurance. But on what basis, then, would anyone think that compulsory vaccination might undermine the expected utility of vaccination policies, if by promoting fairness it gives people motivation to contribute towards the realization of herd immunity? Some authoritative experts in the fields of vaccines and vaccination policies have claimed that coercive vaccination policies could backfire (*Nature* editorial 2018), for example, by undermining people's motivation to vaccinate. However, there is no evidence that current coercive vaccination policies have this effect. For example, when the Italian government introduced vaccination as a requirement for enrolling young children in childcare and primary school, as well as compulsory vaccination for school-age children, child vaccination uptake significantly increased in 2017, within just one year of the policy's introduction. For children born in 2015, MMR vaccine uptake increased by 4.42% (from about 87% to about 92%) and the 6-in-1 vaccine uptake increased from about 93% to about 94.5%, allowing to reach the threshold of herd immunity in 11 of the 20 Italian regions. For children born in 2014, the coverage rate of the 6-in-1 vaccine increased from about 93% to about 95%, and the MMR vaccine coverage rate had an astonishing increase of more than 5%, from about 87% to about 92% (all the data are available from Italian Ministry of Health 2018). It might be too early to draw any definitive conclusion about the effectiveness of such coercive policies: the trend in the next years might be the same as the trend in the last year, or it might be the exact opposite. And the effectiveness of compulsory vaccination might turn out to be a short-run phenomenon, as can sometimes be the case with mandatory vaccination (as we have seen in Chap. 3). Time will tell. Also, of course, it might well be that in some other places a compulsory vaccination policy like the one implemented in Italy, with a 500 euro fine for non-vaccination, would actually backfire. But whether and for how long compulsory vaccination would be effective is more likely to depend on the magnitude of the legal penalty and on the way it is implemented than on the element of compulsion itself. Even if a certain coercive vaccination policy, and compulsory vaccination in particular, were to somehow backfire, this would hardly be an argument against coercion or compulsion. Instead, it would be an argument in favour of enhancing controls on compliance with the legal requirement and on increasing the penalty for non-vaccination in order to contrast the potential backfiring effect. No matter how strongly opposed to vaccination a person is, there will always be a penalty they are not prepared to pay for non-vaccination. And the same consideration applies in the case of people who do not vaccinate out of concerns about safety and effectiveness of vaccines.

This is not to say that we should not try to address people's concerns about vaccines through other policies, such as appropriate information campaigns or strategies aimed at promoting trust between doctors and citizens. Of course, compulsory vaccination would have to be accompanied by policies aimed at increasing people's confidence in vaccines, so that they would be more likely to comply with legal requirements. But we do need to ensure that everybody contributes their fair share to the achievement of the preservation of the public good of herd immunity; and to this end, compulsory vaccination is the ethically acceptable and indeed ethically obligatory policy measure to adopt. It is ethically acceptable because it only compels individuals to fulfil an ethical obligation they independently have (as discussed in Chap. 2), and not to do something supererogatory or beyond the "call of duty". And it is ethically obligatory because fairness itself ought to be an aim of vaccination policies, and fairness demands that everybody take on themselves an equal share of the burdens entailed by the realization of herd immunity; it is unlikely that anything less than compulsion would get closer to this goal. At the same time, compulsory vaccination also fulfils the ethical requirements of expected utility, of fairness, and of the least restrictive alternative.

Let us briefly review in what ways the ethical case for compulsory vaccination complies with the three fundamental ethical requirements of public policies. As for expected utility, compulsory vaccination is likely to be highly effective at achieving the stated goal, in that it would allow for the highest possible vaccination rates compared to any other policy (with the exception of forced vaccination), provided the legal penalty is sufficiently burdensome and adequately implemented. As for fairness, compulsory vaccination would distribute the burdens of such collective enterprise in the fairest possible way, again with the exception of forced vaccination. As for liberty, compulsory vaccination does not infringe on any liberty right not to be vaccinated, because such rights do not exist in this context.

One might reasonably observe that there is a difference between adult and child vaccination. For instance, parents do pay taxes and thus make personal sacrifices for the sake of the collective good, but children are obviously not required to make an equivalent contribution, given that they do not have the resources or the capacities to do it (indeed, they are on the receiving, as opposed to contributing, end of important goods). So how do the requirements of fairness, according to which each individual ought to make their fair contribution to herd immunity, apply to children? We obviously do not expect those who do not have the resources or the capacities—such as children, but also those who live in extreme poverty—to make their contribution to public goods through taxation. The legal obligation to make one's contribution obviously presupposes the capacity to make one's contribution, and children, as well as other groups, often lack these capacities. But the case of vaccination is different. Since children do have the capacity to make their contribution by being vaccinated, assuming that vaccination does not pose any significant cost on them, it follows that they should be subject to the same legal obligation. It might be more problematic to say that they have a *moral* obligation, since on many plausible accounts of moral responsibility, being an autonomous agent is a necessary for condition for being subjects of moral responsibility, and very young children do not count as autonomous agents. But if, as we saw in Chap. 2, the moral responsibility is supposed to fall on parents, who make choices on behalf of their children, then the fact that children cannot have moral obligations does not pose any problem; ultimately, it is parents who have both the moral and the legal responsibility to vaccinate their children.

On Non-medical Exemptions from Compulsory Vaccination

What I have said so fair about the fairness-based moral obligation to vaccinate oneself or one's children, and about the fairness-based justification for compulsory vaccination policies, has implications for the issue of conscientious objection to vaccination. Non-medical vaccine exemptions to vaccination mandates are granted in most—though not all—US states, where vaccination is mandatory for enrolling children in school. It is not surprising that, within the debate on the ethics of vaccination and of vaccination policy, conscientious objection to vaccination has received quite a lot attention in the last few years (see, e.g., Clarke et al. 2017; Navin and Largent 2017; Giubilini et al. 2017). After all, within medical ethics more broadly, conscientious objection and appeals to "conscience" have become the standard way of claiming an alleged right to stick to one's own allegedly ethical preferences where one would normally be required—either by professional or legal obligations, or both—to do otherwise. Appeals to conscience are often deployed to suggest the idea that there can be reasonable moral disagreement about certain issues and that, in virtue of this reasonable moral disagreement, people should be free to follow their conscience. Conscience is like a magic word that confers unwarranted authority to any view or idea one might hold (Giubilini 2016).

Conscientious objection to vaccination is regulated differently in different US states. (It is worth noting that, in the context of US policy, we are talking about exemption to mandatory rather than to compulsory vaccination, but the same ethical considerations apply.) Sometimes non-medical exemptions are granted only for religious reasons, but not for other kinds of personal reasons. This is currently the case in 29 US states; Vermont in 2016 was the last state, in chronological order, to implement what Navin and Largent have dubbed the "prioritizing religion" model (Navin and Largent 2017). This model, however, is marred by a number of grave practical and ethical shortcomings: not only is it very difficult to define what counts as "religious" and to assess whether people's views are religious, but more importantly, when it comes to protecting individual liberties, privileging religion over other views fails to promote that minimum level of neutrality among different conceptions of the good that liberal societies are supposed to promote (Navin and Largent 2017). Hence, I will refrain from any further discussion of this model.

There are different possible views on how conscientious objection to vaccination, whether compulsory or mandatory, should be regulated. Some have observed that in the same way as conscientious objection is normally granted in other contexts on the basis of deeply held ethical or religious views, for example, in the case of pacifists' exemptions from conscription, so those with ethical or religious beliefs against vaccination ought to be exempted from vaccination mandates, provided there is no significant risk for third parties (Navin and Largent 2017). In order to keep the number of conscientious objections to a minimum, and thus ensure that herd immunity is realized while individual freedom remains reasonably protected, Navin and Largent have endorsed what they call the "inconvenience" model, as recently implemented, for example, in Michigan: in brief, parents applying for non-medical exemptions to vaccination mandates are required to go through some burdensome and inconvenient procedures, including not only bureaucratic procedures but also things like attending immunization education sessions. In this way, only those who have genuine ethical or religious objections to vaccines are likely to be willing to go through such procedures, and therefore this model is likely to strike a reasonable balance between protection of individual freedom (of those with genuine ethical or religious objection to vaccines) and effectiveness in reaching sufficiently high vaccination rates. And in fact, there is evidence in support of Navin and Largent's point, since burdensome exemption rates have been shown to lower exemption requests (Blank et al. 2013; Omer et al. 2012; Rota et al. 2001). For example, in Michigan, one year after the implementation of the inconvenience model in 2106, exemption rates fell by 39% (and by 60% in the Detroit area) (Higgins 2016; Navin and Largent 2017).

A second approach is slightly less favourable to considerations of individual liberty, but still aims to strike a balance between liberty, effectiveness of vaccination policies, and fairness. According to this approach, individuals who are exempted from vaccination mandates should be required to compensate society for their failure to contribute to herd immunity. In this way, people would still be free to refuse vaccination, but they would be discouraged from doing so, and would have to pay a price if they did refuse. There are two ways of understanding this requirement. Both interpretations are based on a comparison between conscientious objection to vaccination and conscientious objection to the military service. According to one version of this view, as we saw in Chap. 3, individuals who refuse vaccination should be required to make an alternative contribution to society in the form of a tax, in lieu of their contribution to herd immunity through vaccination. The size of this alternative contribution should depend on the extent to which non-vaccination is likely to harm others, which in turn depends on factors such as the severity and morbidity of the disease in question (Clarke et al. 2017). However, while this approach points in the right direction by giving *some* weight to considerations of fairness, there are two problems with it. The first is that any case of non-vaccination is unlikely to harm others where herd immunity exists, and therefore people would be required to make an alternative contribution only where vaccination rates are low, which would contradict a requirement of fairness. Second, people who are exempted from military service are typically not taxed, but are required to make an alternative contribution to society. But a tax in this context would look a lot like a financial penalty, rather than an alternative contribution. Overall, the problem with this approach is that, while it is similar to my suggested model of compulsory vaccination in that it proposes a financial penalty for non-vaccination, it only applies this tax/penalty in proportion to the actual risk posed by non-vaccination. The implication is that if the risk is negligible, vaccine refusers could simply get away with non-vaccination and free-ride on herd immunity, failing to make their fair contribution to the fulfilment of the collective obligation to realize herd immunity.

In order to fulfil a requirement of fairness and to be consistent with other cases of conscientious objection (in particular towards military service), we might envision a second variant of the same approach, wherein those who refuse vaccination should make some alternative contribution to society—one that is independent of the actual risk that a specific case of non-vaccination poses on others and that is roughly equivalent to the contribution one would make to public health by being vaccinated. For example, in the same way as those who refuse to enrol in the military are typically required to provide alternative *social* services (such as working in public libraries, assisting with provision of services for people with disabilities, etc.), people who refuse vaccination could be required to make some alternative contribution to public health, such as preparing healthy meals for children at school or participating in fundraising activities in support of infectious disease research. This is what has been called the "contribution model" for non-medical vaccine exemptions (Giubilini et al. 2017): the point here is that those who refuse vaccination should not be "punished", but rather required to make their fair contribution to the public good in question, or to a roughly equivalent good, in other ways. While this model of dealing with vaccine refusal is preferable to the ones described so far in that it goes further than the alternative ones in promoting liberty, expected utility, and fairness, it does have its own shortcomings. For how are we to determine this rough equivalence? The problem is that the rough equivalence between the two types of contribution cannot be measured in terms of the actual impact of individual behaviour on the good in question, since in either case such contribution is likely to be negligible. So the equivalence needs to be measured in terms of the societal benefit of the (public) goods to which one is required to contribute. But the problem is that the alternative public health goods contemplated by the contribution model are not really commensurable with herd immunity: herd immunity is very likely to prevent illness and even death from vaccine-preventable infectious diseases, an outcome that would be extremely beneficial to many individuals and to society. The goods promoted by the other suggested contributions to public health are either less beneficial (such as giving children healthy food) or of more uncertain and of longer-term realization (funding research to find cures for dangerous infectious diseases). What we want when implementing vaccination policies is the realization of herd immunity through a fair distribution of the burdens entailed *by herd immunity itself.* None of the alternative contributions proposed would go in this direction. Therefore, none of the alternatives would promote fair distribution of the burdens entailed by the realization of herd immunity, although they would promote some form of fairness.

On the basis of the argument provided in this chapter, it should be apparent to those who agree with it that the problem with all these proposals is that, even if they are consistent with or would promote the realization of herd immunity, they assume that herd immunity is the *only* legitimate aim of vaccination policies. In other words, they disregard the fundamental value of fairness as an end in itself of morally acceptable policies aimed at protecting important public goods, which I have defended in this chapter. Simply put, and as an obvious implication of my argument in this book, fairness in the distribution of the burdens of herd immunity is itself an end of vaccination policies, and therefore granting some people non-medical exemptions would violate the fairness requirement.

But what about the analogy with conscientious objection to conscription? After all, one might insist, if we accept that pacifists should be exempted from contributing to a public good like national security—and that they should at most be required to provide some alternative service to society—why should we not adopt the same type of exemption policy in the case of vaccination requirements? I offer two very short answers to this reasonable observation. First, the analogy with conscientious objection to military service is just one possible analogy. An alternative analogy—the one on which I have sometimes relied on here—is the one with taxation. In the case of taxation, exemptions are normally not granted, not even by requesting individuals to make alternative contributions instead. If someone thinks that the choice between the two analogies is merely arbitrary, and therefore that there are no particular reasons for preferring the taxation analogy over the military analogy, they would have to acknowledge that there are no particular reasons for preferring the military analogy over the taxation analogy either; therefore, the taxation analogy is not inferior to the military analogy, and my preference for the analogy with the taxation model cannot really be criticized. Second, not only is the taxation analogy not inferior to the military analogy, but, I would argue, it is actually preferable: requesting each person to pay their fair share of taxes would ensure that certain public goods are realized through a fair distribution of the burdens they entail—which, as I have argued in this chapter, should be the final aim of ethically acceptable vaccination policies; on the contrary, allowing conscientious objection in the military context would not guarantee that the public good of national defence is realized *fairly*.

But does then my argument imply that conscientious objection to conscription is also ethically wrong and that for the sake of fairness pacifists should be forced to enrol in the military where conscription is in place? And would I be prepared to accept this implication in order to defend my argument for compulsory vaccination without the possibility of non-medical exemptions? As unappealing and perhaps counterintuitive as this position might sound, the argument I have developed in this chapter, and indeed throughout the entire book, compels me to give an affirmative answer to both questions. I am happy to accept this implication. There should be compulsory vaccination policies in place that do not allow for any non-medical exemptions, and there should be no conscience-based exemption to conscription, at least if we agree that national defence should be guaranteed also through conscription and not only through professional soldiers; if we thought that national defence ought to be protected only by professionals in the military, the analogy with vaccination would not apply, since there is no equivalent profession in the case of vaccination (not least because we would need to pay a huge number of "professionals" to achieve herd immunity, way more than those required for national security).

Conclusion

As I have argued in this book, an ethical approach to vaccination choices and to vaccination policy supports the following claims:

1. There is a collective responsibility, or collective moral obligation, to realize herd immunity.
2. There is an individual moral obligation to contribute to the realization of herd immunity by being vaccinated or by vaccinating one's children.
3. The state has an institutional responsibility to implement vaccination policies that can guarantee at least the realization of herd immunity.
4. If the aim of vaccination policies were merely the realization of herd immunity, then a principle of least restrictive alternative would imply that the state has an institutional responsibility to implement the least restrictive policy that would be effective in achieving this goal.
5. However, a principle of fairness requires that everybody—not just the smallest number of people that can realize herd immunity—makes their fair contribution to herd immunity by getting vaccinated.

6. The existence of an individual obligation to be vaccinated or to vaccinate one's children implies that the state is morally justified in requiring each individual to be vaccinated or to vaccinate their children, in the absence of legitimate medical reasons for exemptions; in other words, compulsory vaccination without non-medical exemptions is ethically justified.

7. A principle of fairness in the distribution of the burdens entailed by an important public good like herd immunity implies that the state ought to require each individual to be vaccinated or to vaccinate their children, in the absence of legitimate medical reasons for exemptions; in other words, enforcing compulsory vaccination without non-medical exemptions is an ethical obligation of states.

8. Compulsory vaccination meets the requirements posed by the ethical principles that should inform policymaking, namely, maximization of expected utility, fairness, and least restrictive alternative, if properly understood.

REFERENCES

Bambery, B., Selgelid, M., Maslen, H., Pollard, A., & Savulescu, J. (2013). The Case for Mandatory Flu Vaccination. *American Journal of Bioethics, 13*(9), 38–40.

Beard, F. H., Leask, J., & McIntyre, P. B. (2017). No Jab, No Pay and Vaccine Refusal in Australia: The Jury Is Out. *The Medical Journal of Australia, 206*(9), 381–383.

Bester, J. C. (2015). Vaccine Refusal and Trust: The Trouble with Coercion and Education and Suggestions for a Cure. *Journal of Bioethical Inquiry, 12,* 555–559.

Blank, N. R., Caplan, A. L., & Constable, C. (2013). Exempting Schoolchildren from Immunizations: States with Few Barriers Had Highest Rates of Nonmedical Exemptions. *Health Affairs, 32*(7), 1282–1290.

Brennan, J. (2018). A Libertarian Case for Mandatory Vaccination. *Journal of Medical Ethics, 44,* 37–43.

Camerer, C., & Thaler, R. H. (1995). Anomalies: Ultimatums, Dictators, and Manner. *Journal of Economic Perspectives, 9,* 209–219.

Clarke, S., Giubilini, A., & Walker, M. J. (2017). Conscientious Objection to Vaccination. *Bioethics, 31*(3), 155–161.

Dawson, A. (2007). Herd Protection as a Public Good: Vaccination and Our Obligations to Others. In A. Dawson & M. Verweij (Eds.), *Ethics, Prevention, and Public Health* (pp. 160–178). Oxford: Clarendon Press.

Flanigan, J. (2014). A Defense of Compulsory Vaccination. *HEC Forum*, 26, 5–25.

Giubilini, A. (2014). The Paradox of Conscientious Objection and the Anemic Concept of Conscience. Downplaying the Role of Moral Integrity in Healthcare. *Kennedy Institute of Ethics Journal*, 24(2), 159–185.

Giubilini, A. (2016). Conscience. In E. N. Zalta (Ed.), *The Stanford Encyclopedia of Philosophy* (Winter 2016 ed.). Palo Alto: Stanford University Press.

Giubilini, A., Douglas, T., & Savulescu, J. (2017). Liberty, Fairness, and the 'Contribution Model' for Non-medical Vaccine Exemption Policies: A Reply to Navin and Largent. *Public Health Ethics*, 10(3), 235–240.

Giubilini, A., Douglas, T., & Savulescu, J. (2018). The Moral Obligation to Be Vaccinated: Utilitarianism, Contractualism, and Collective Easy Rescue. *Medicine, Health Care, and Philosophy*. https://doi.org/10.1007/s11019-018-9829-y. Online first 10 Feb.

Guth, W., Schmittberger, R., & Schwarze, B. (1982). An Experimental Analysis of Ultimatum Bargaining. *Journal of Economic Behavior and Organization, 3*, 367–388.

Higgins, L. (2016, January 28). More Michigan Parents Willing to Vaccinate Kids. *Detroit Free Press*. Retrieved April 30, 2018, from http://www.freep. com/story/news/education/2016/01/28/immunization-waivers-plummet-40-michigan/79427752/

Italian Ministry of Health. (2018). *I dati nazionali al 2017 sulle coperture vaccinali dell'eta' pediatrica e dell'adolescente*. Available from http://www.salute. gov.it/portale/news/p3_2_1_1_1.jsp?lingua=italiano&menu=notizie&p=dal ministero&id=3348. Last accessed 4 May 2018.

Mill, J. S. (1859). *On Liberty*. London: Walter Scott Publishing. In *The Project Gutenberg Ebook of On Liberty*. Available at https://www.gutenberg.org/ files/34901/34901-h/34901-h.htm

Nature. (2018). Laws Are Not the Only Way to Boost Immunization. Editorial Article. *Nature, 553*, 249–250.

Navin, M. (2015). *Values and Vaccine Refusal: Hard Questions in Ethics, Epistemology, and Health Care*. New York: Routledge.

Navin, M. C., & Largent, M. A. (2017). Improving Nonmedical Vaccine Exemption Policies: Three Case Studies. *Public Health Ethics*, 10(3), 225–234.

Omer, S. B., et al. (2012). Vaccination Policies and Rates of Exemption from Immunization, 2005–2011. *New England Journal of Medicine, 367*, 1170–1171.

Pierik, R. (2016). Mandatory Vaccination: An Unqualified Defence. *Journal of Applied Philosophy*. https://doi.org/10.1111/japp.12215.

Rawls, J. (1971/1999). *A Theory of Justice* (Rev. ed.). Cambridge, MA: Harvard University Press.

Rota, J. S., Salmon, D. A., Rodewald, L. E., Chen, R. T., Hibbs, B. F., & Gangarosa, E. J. (2001). Processes for Obtaining Nonmedical Exemptions to State Immunization Laws. *American Journal of Public Health, 91*(4), 645–648.

Salmon, D. A., & Siegel, A. W. (2001). Religious and Philosophical Exemptions from Vaccination Requirements and Lessons Learned from Conscientious Objectors from Conscription. *Public Health Reports, 116*(4), 289–295.

Selgelid, M. (2009). A Moderate Pluralist Approach to Public Health Policy and Ethics. *Public Health Ethics, 2*(2), 195–205.

Tabibnia, G., & Lieberman, M. D. (2007). Fairness and Cooperation Are Rewarding. *Annals of the New York Academy of Sciences, 1118*(1), 90–101.

Tabibnia, G., Satpute, A. B., & Lieberman, M. D. (2008). The Sunny Side of Fairness: Preference for Fairness Activates Reward Circuitry (and Disregarding Unfairness Activates Self-Control Circuitry). *Psychological Science, 19*(4), 339–347.

Wang, E., Clymer, J., Davis-Hayes, C., & Buttenheim, A. (2014). Nonmedical Exemptions from School Immunization Requirements: A Systematic Review. *American Journal of Public Health, 104*(11), e62–e84.

INDEX[1]

[1] Note: Page number followed by 'n' refer to notes.

.